✦ HORSEKEEPING SKILLS LIBRARY ✦

Horse Handling & Grooming

A STEP-BY-STEP PHOTOGRAPHIC GUIDE TO MASTERING OVER 100 HORSEKEEPING SKILLS

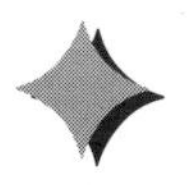

CHERRY HILL

PHOTOGRAPHY BY RICHARD KLIMESH

Storey Publishing

The mission of Storey Publishing is to serve our customers by publishing practical information that encourages personal independence in harmony with the environment.

Edited by Elizabeth McHale
Cover design by Eugenie Delany
Text design by Cynthia McFarland
Production by Therese Lenz and Allyson L. Hayes
Line drawings designed by Cherry Hill and drawn by Elayne Sears
Indexed by Susan Olason

Printed in the United States by Versa Press

20 19 18

LIBRARY OF CONGRESS CATALOGING-IN-PUBLICATION DATA

Hill, Cherry, 1947–
Horse handling and grooming : a step-by-step photographic guide to mastering over 100 horsekeeping skills / Cherry Hill : with photographs by Richard Klimesh.
p. cm. — (Horsekeeping skills)
Includes bibliographical references and index.
ISBN-13: 978-0-88266-956-4 (pb : alk. paper)
1. Horses—Handling. 2. Horse—Grooming. 3. Horses—Handling—Pictorial works.
4. Horse—Grooming—Pictorial works.
I. Title. II. Series.
SF285.6.H54 1997
636.1'0833—dc20 96-30623
CIP

Contents

Dedication

To Cindy Foley

Other Books by Cherry Hill

101 Arena Exercises

Horsekeeping on a Small Acreage

Becoming an Effective Rider

Your Pony, Your Horse

From the Center of the Ring

The Formative Years

Making Not Breaking

Maximum Hoof Power

Horse for Sale

Horse Health Care

Acknowledgments

Special thanks to my husband and partner, Richard Klimesh, for his sense of humor and excellent help with the photographs.

Thanks to Sue DeGrazia for being a photo model. Also, thanks to my horses Zinger, Sassy, Zipper, Dickens, Blue, Aria, Seeker, and Drifter for their patience and cooperation as photo models.

I also extend my gratitude to the following companies for supplying products used in this book: **Ariat International,** for safe and comfortable boots for our human models; **BMB Animal Apparel,** for attractive, safe halters, sheets, and blankets; **Wahl Clipper Company,** for Stable Pro and rechargeable clippers; **Metropolitan Vacuum Cleaner Company** for Vac' N' Blo horse vacuum; and **Home Impressions** for Groomex Vacuum.

Preface

We sometimes run into Murphy's Law at our place when one of us is demonstrating how to do something for a student, photo, or video. We often laugh and say "Watch me tie this knot" before we begin a photo shoot or demonstration. That's because once, many years ago, I was demonstrating for a horse-training class how to tie a bowline knot around a horse's neck. I had tied many a bowline knot before, but that day I wanted to use a technique I had recently seen (but not practiced) that I thought would make teaching the knot much easier. In order for the group to be able to see what I was doing, I had to position the horse differently from how I would if I were just tying the knot as a matter of course. Also, so I wouldn't block the students' view, I had to stand off to the left, almost at the horse's head, instead of at the horse's shoulder. When I finished my demonstration of this ordinarily simple little knot, the result was definitely not a bowline.

Trying to incorporate a new technique while standing in an unfamiliar position was not a good idea. And to make matters worse, some of my students had been committing to memory the faulty procedure I had demonstrated.

You've probably already figured out one moral of the story: Stick to tried-and-true methods. But I have other reasons for telling this anecdote. Each task or skill involved with horses should be approached as a formal lesson for both you and your horse. You will have much better results if you separate a task into its components and master each part, paying attention to every detail.

When it comes to caring for and handling horses, there are many ways to do things. Usually there is one best way, several acceptable variations, and some wrong ways! Find the best way to perform each skill and practice it that way. Then, as you become familiar with a skill, you might want to vary the way you perform it according to your height, hand size, facilities, tack, equipment, and experience. Once you have customized a skill to fit you and your horse, establish a consistent routine for performing it each time. Horses are creatures of habit, and they feel confident and content when their handlers behave consistently and predictably. Following a routine will also result in a greater chance for efficiency and effectiveness.

Finally, if you keep an open mind and a sense of humor about what you're doing, you'll find that everyday tasks will be enjoyable and rewarding by themselves. In fact, I'm looking forward to giving Seeker a thorough grooming and touch-up clip before her lesson today. See you down at the barn!

Safe Handling and Housing

Equipment for the Handler

Always use strong, well-made, and well-fitted equipment. Although safety helmets are usually associated with riding, there are times when you are handling horses from the ground that it is a good idea to wear one. Protect your feet by wearing well-made, sturdy boots. Whenever possible wear gloves, especially when handling ropes. And always use safe horse-handling techniques.

▲
Sturdy Boots
Boots should have heels, good traction, and, if possible, an extra piece of leather sewn across the toe. These boots have a toe cap, which provides extra protection if a horse should step on your foot.

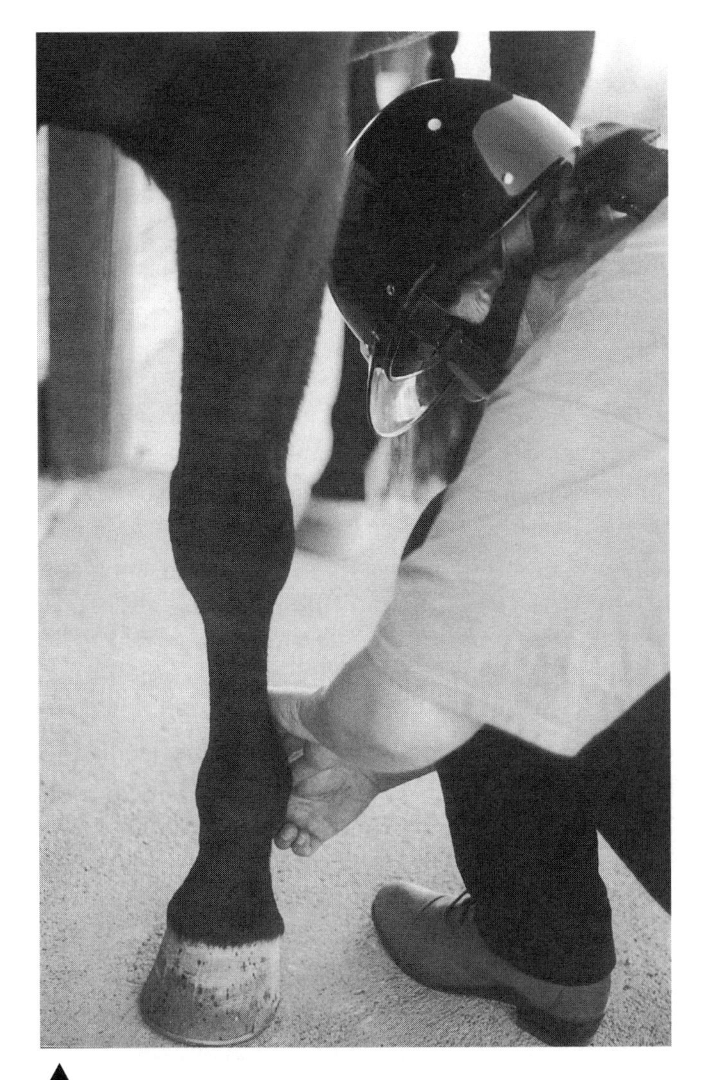

▲
Safety Helmet
If you are inexperienced or you are working with a young or green horse, it would be to your benefit to wear a protective helmet. When you are working on a horse's legs, the horse could accidentally hit you in the head when he stomps at a fly or he could move suddenly and knock you into a wall or fence or onto the ground.

▲
STRONG EQUIPMENT AND LEATHER GLOVES
Always use strong, well-made, well-fitted equipment. Leather gloves will give you a better grip on ropes and protect your skin from the pain of a rope burn if the rope ever zings through your hand unexpectedly.

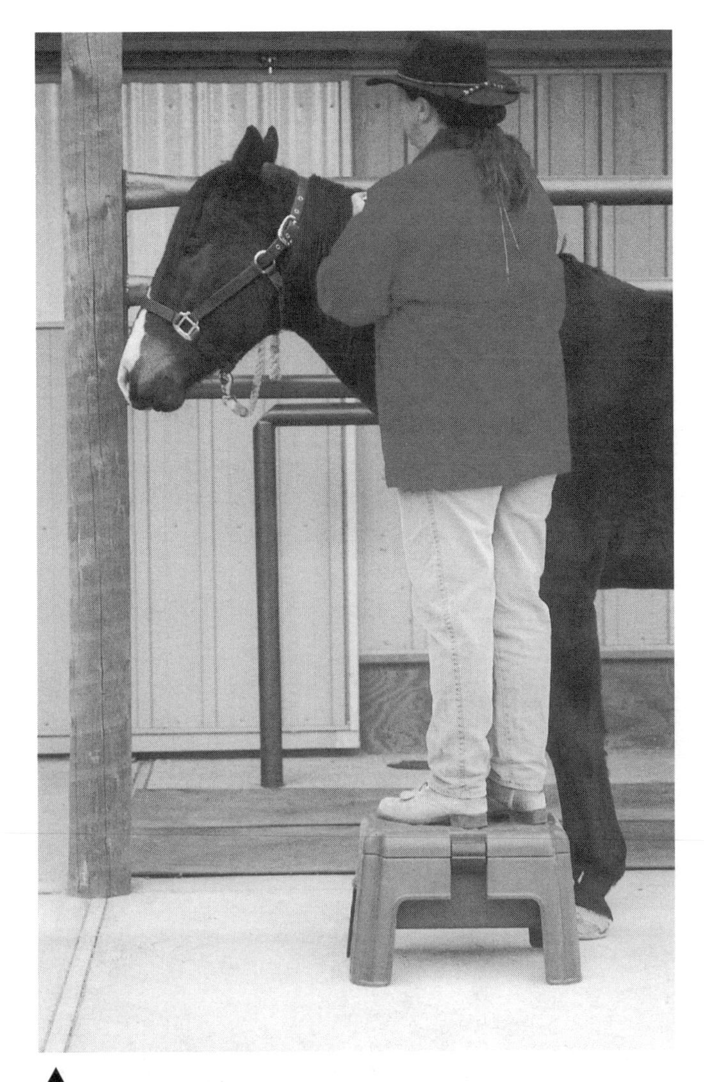

▲
STEP STOOL
When working on your horse's mane, if you need to get a bird's-eye view, use a steady step stool and place it next to the horse, not in front of him or underneath him. Also, if you have long hair, fasten it securely out of the way and wear a hat to further contain it.

Safely Housing a Horse

Whether your horse is housed in a stall or an outdoor pen, follow safe practices and provide safe eating and water areas.

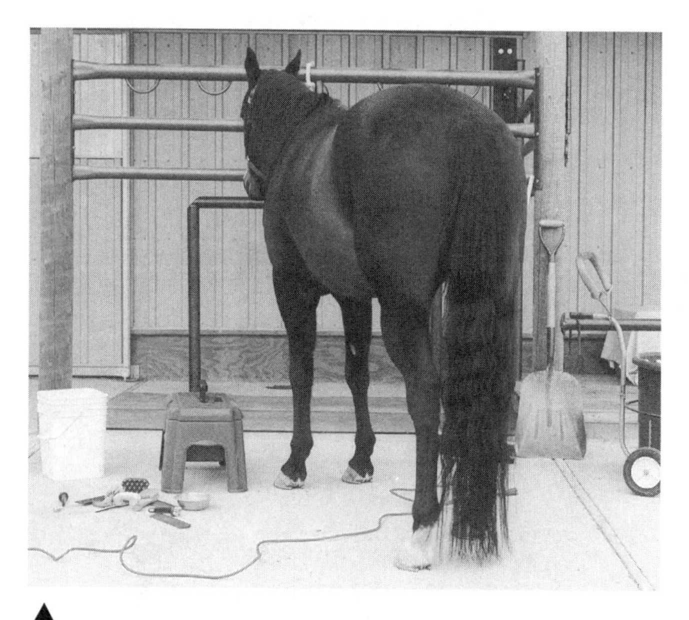

▲
Avoid Clutter
Follow safe practices at all times. Never leave a horse in a dangerously cluttered area like this. He could easily become hurt or damage your equipment.

▲
The Stall
Provide a safe, comfortable place for your horse to live. If you horse lives in a stall part of the time, be sure the stall has plenty of air and light, smooth, safe walls, and a safe feeder and waterer.

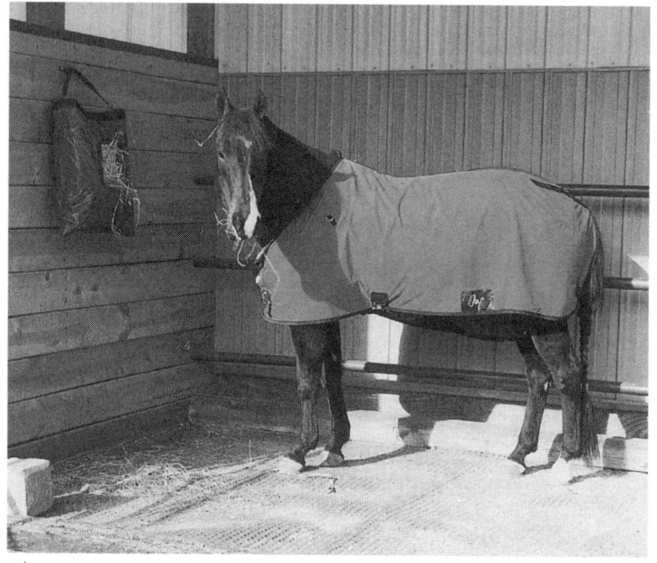

▲
Outdoor Pen
If your horse lives in an outdoor pen, provide him with a clean, safe eating area, shelter, and perhaps a sheet or blanket.

KNOW FIRST AID and have proper supplies and equipment on hand, including Betadine, sterile gauze pads, conforming gauze roll, crepe bandage, scissors, latex gloves, thermometer, and triple antibiotic ointment.

Handling Equipment

HALTER SELECTION

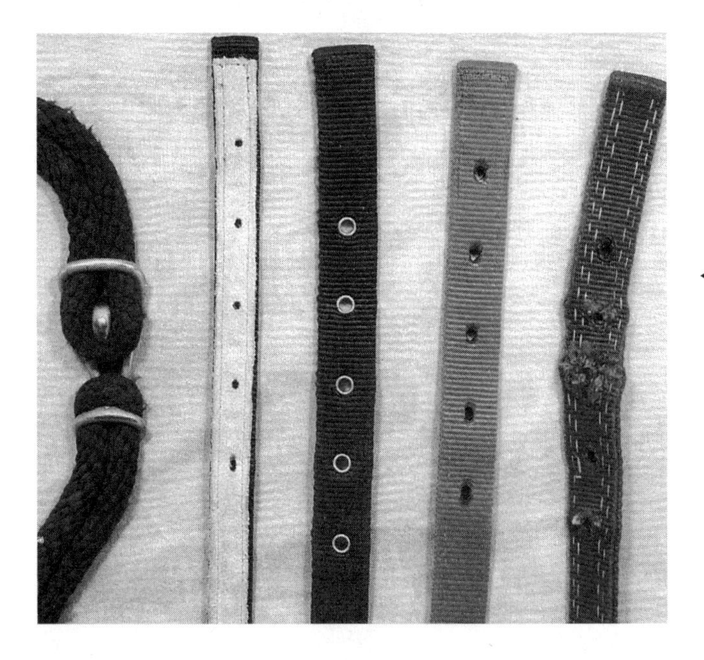

◀ CHOOSE SAFE HALTERS

When choosing a halter, pay attention to the manner in which it buckles. From the left, various halter straps: a rope foal halter with a loop and a hook closure; a weanling halter made from narrow web with leather lining for extra strength; a nylon web halter with holes reinforced with metal grommets; a nylon web halter with customary burned-in holes; and a ragged nylon halter. All halter straps except the last one would be safe to use on a horse.

◀ CHECK THE HARDWARE

Pay attention to the quality of the hardware. The strongest rust-resistant metals for horse hardware are solid brass or bronze, but these are often expensive. Much horse hardware is plated die cast. Die cast is a mixture of non-iron metals that have been heated and poured into a mold. Die-cast hardware that is plated with brass or nickel might be of good quality but if the die-cast material is weak and porous, it could easily snap under pressure. Steel hardware is strong, but once the plating (brass, nickel) wears off, the piece will begin to rust. To determine if a buckle contains iron (or steel, which contains iron), hold a magnet to it. The iron in the metal will attract a magnet.

HALTER FIT

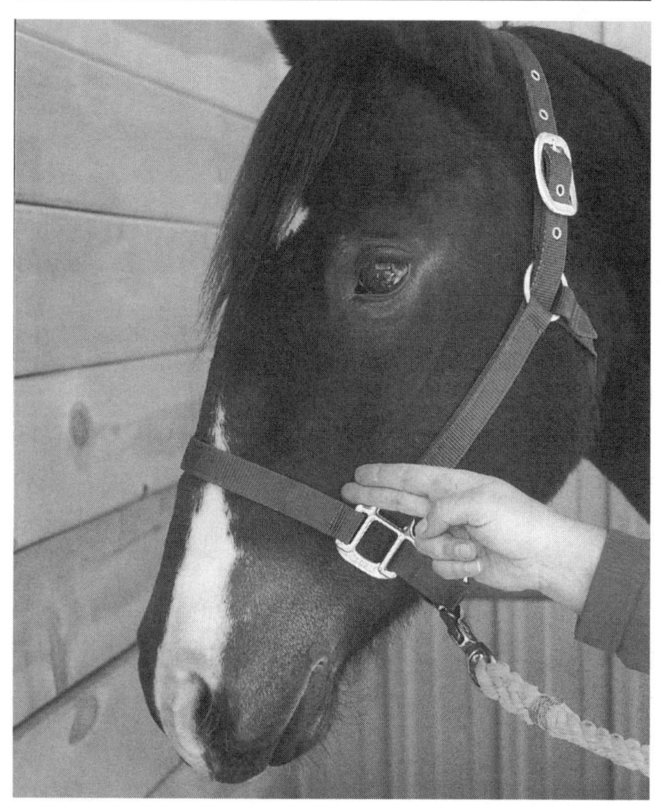

▲
GET A PROPER FIT
This shows a properly fitted halter with the noseband resting two fingers below the prominent cheekbone.

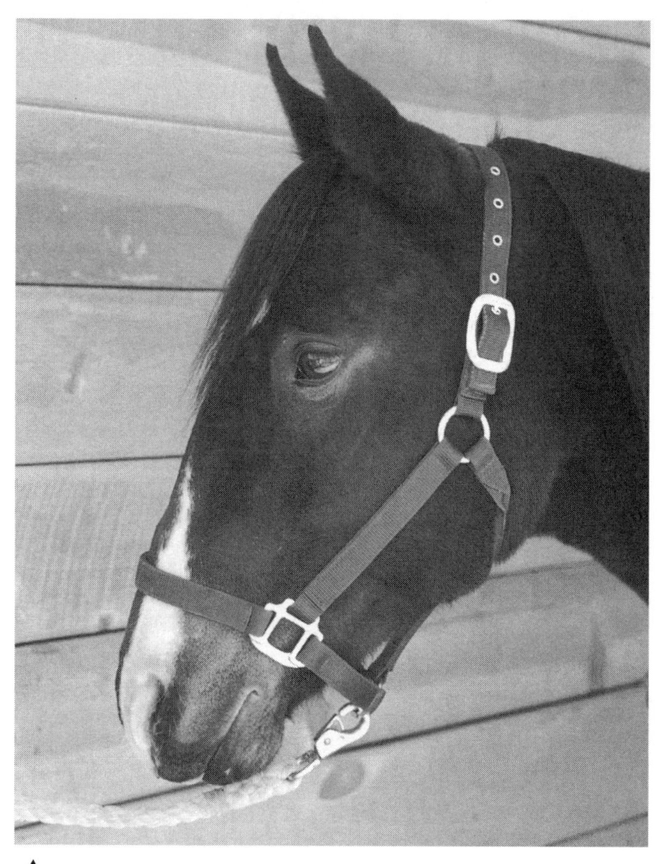

▲
PROBLEM FIT: BUCKLED TOO LOW
Here is the same halter buckled too low on the horse's nose. The noseband is lying on the fragile tip of the nasal bone, which could fracture if too much pressure is exerted on the noseband.

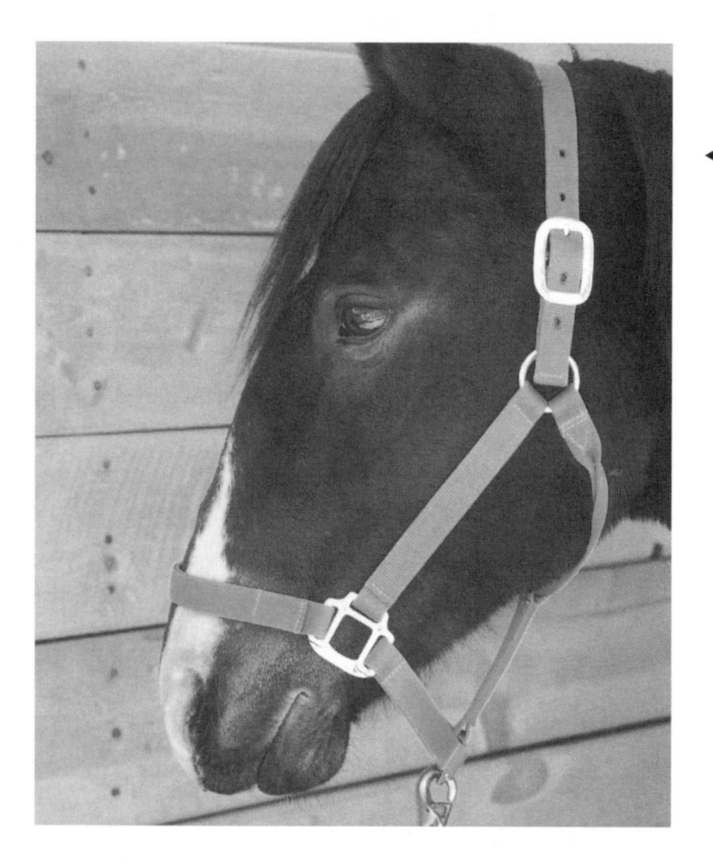

◀ PROBLEM FIT: HALTER TOO LARGE
This halter is not only adjusted too low it is also too large, as evidenced by the huge noseband and drooping throatlatch.

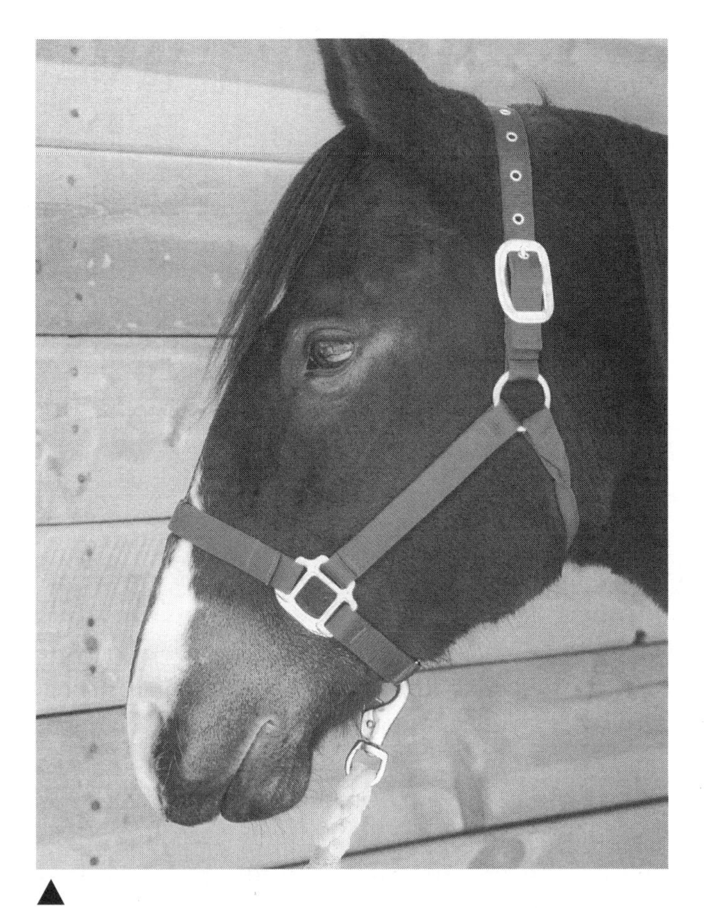

▲
PROBLEM FIT: HALTER TOO SMALL

This halter is too small and tight in the noseband and throatlatch, putting constant pressure on the poll and nose and making it difficult for the horse to move his jaw or swallow properly. This can "deaden" the tissues where there is pressure, resulting in a loss of response to cues.

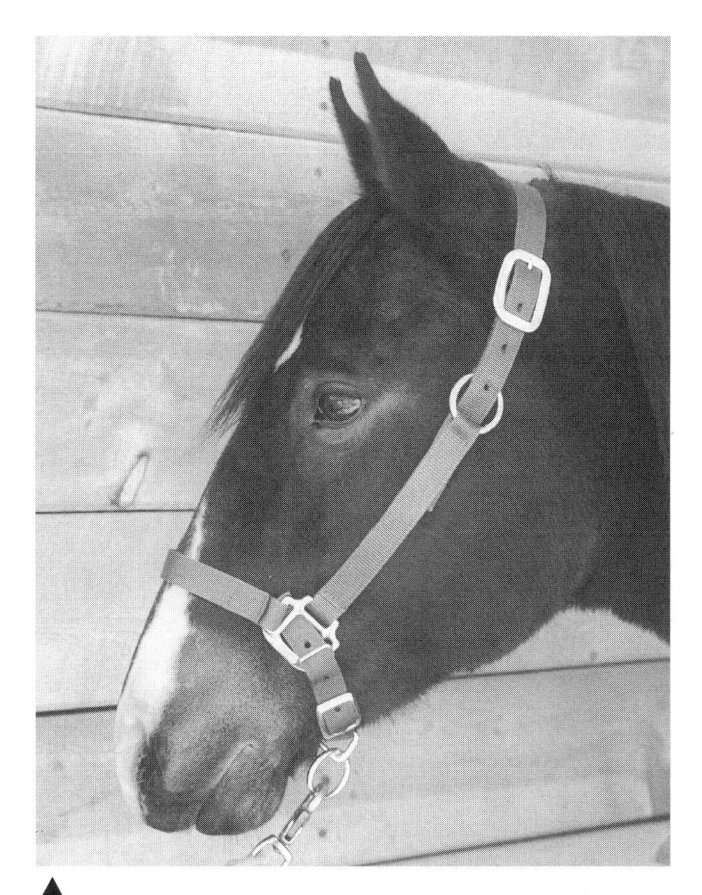

▲
GROOMING HALTER

This is a properly fitted grooming halter. Note that this halter does not have a throatlatch, making it easier to groom the throat and cheek areas. Also, the noseband is adjustable so you can unbuckle it when you want to groom or clip the chin.

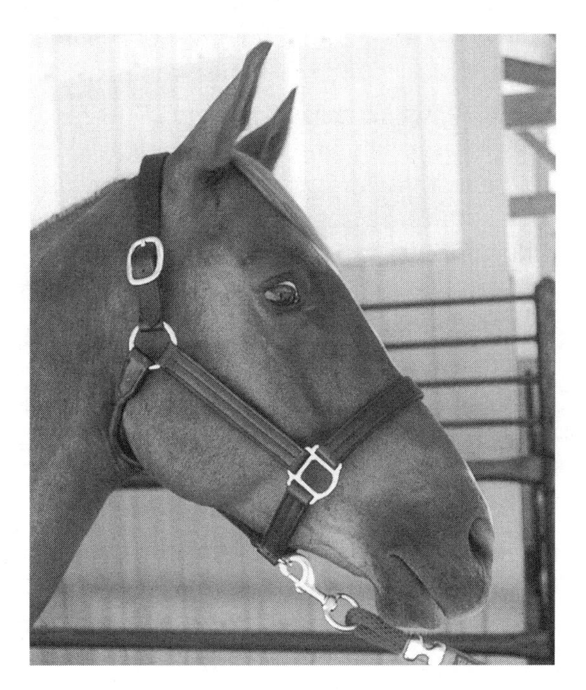

◀ LEATHER HALTER

This is a very strong, well-made, properly fitted leather halter with crownpiece adjustments on both the near side and off side. Leather halters are excellent for promotional photos and in-hand work. Some leather halters are not strong enough to use for tying because leather, especially once it has been exposed to sweat, dirt, and sun, may break when a horse pulls back.

Choosing a Lead Rope

◀ Choose a Good Lead Rope

This is a good lead rope: 10 feet of ⅝-inch cotton rope, back spliced at both ends with the ends of the splice whipped with cotton cord. A nonrusting, heavy-duty bull snap has been spliced into one end. Because cotton is a soft natural fiber, it is the least abrasive, so it causes fewer rope burns than will other types of rope. However, cotton rope tends to fray. To prevent this and to extend the life of your cotton ropes, keep them clean and dry.

Types of Rope

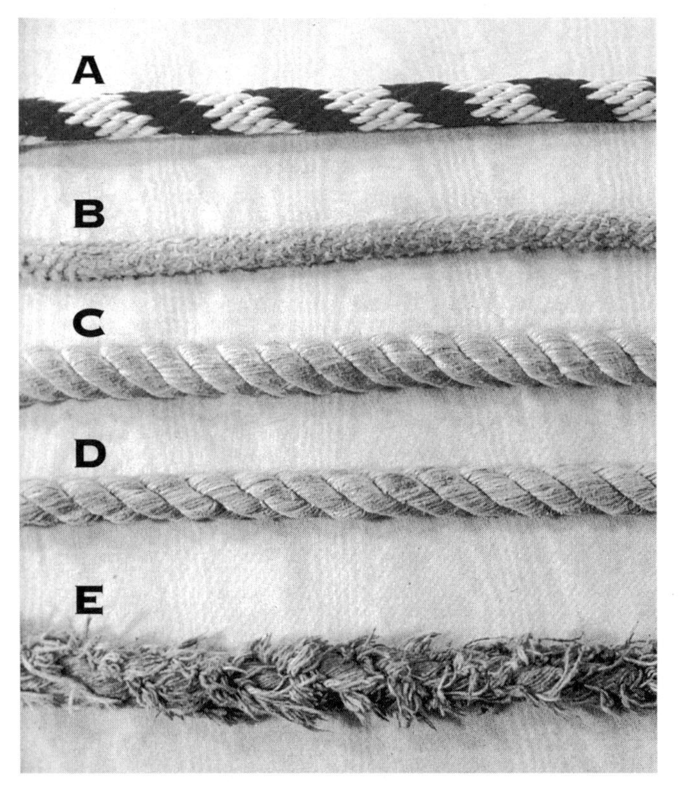

A **½" soft nylon rope.** This is a synthetic fiber that resists rotting or fraying. It is more than four times as strong as a cotton rope of the same diameter. It is nearly as soft as cotton, but its smooth surface can be difficult to grip and may cause rope burn.

B **⅜" hard, nylon, lariat-type rope.** This rope is strong and resists rotting and fraying, but it is too narrow to hold easily. It can cause severe rope burns due to its slick surface, small diameter, and stiffness, and it is difficult to tie.

C **¾" cotton rope.** Cotton rope is strong when kept dry and clean, but it will rot when damp or dirty. This is the easiest rope to grip with gloves, but it may be too thick for a child's hand.

D **⅝" cotton rope.** Because of its smaller diameter, this rope might be more comfortable for a small woman's or child's hand.

E **⅝" cotton rope.** This rope has been allowed to become dirty and wet. The cotton fibers have frayed and broken to such an extent that this rope is no longer safe to use.

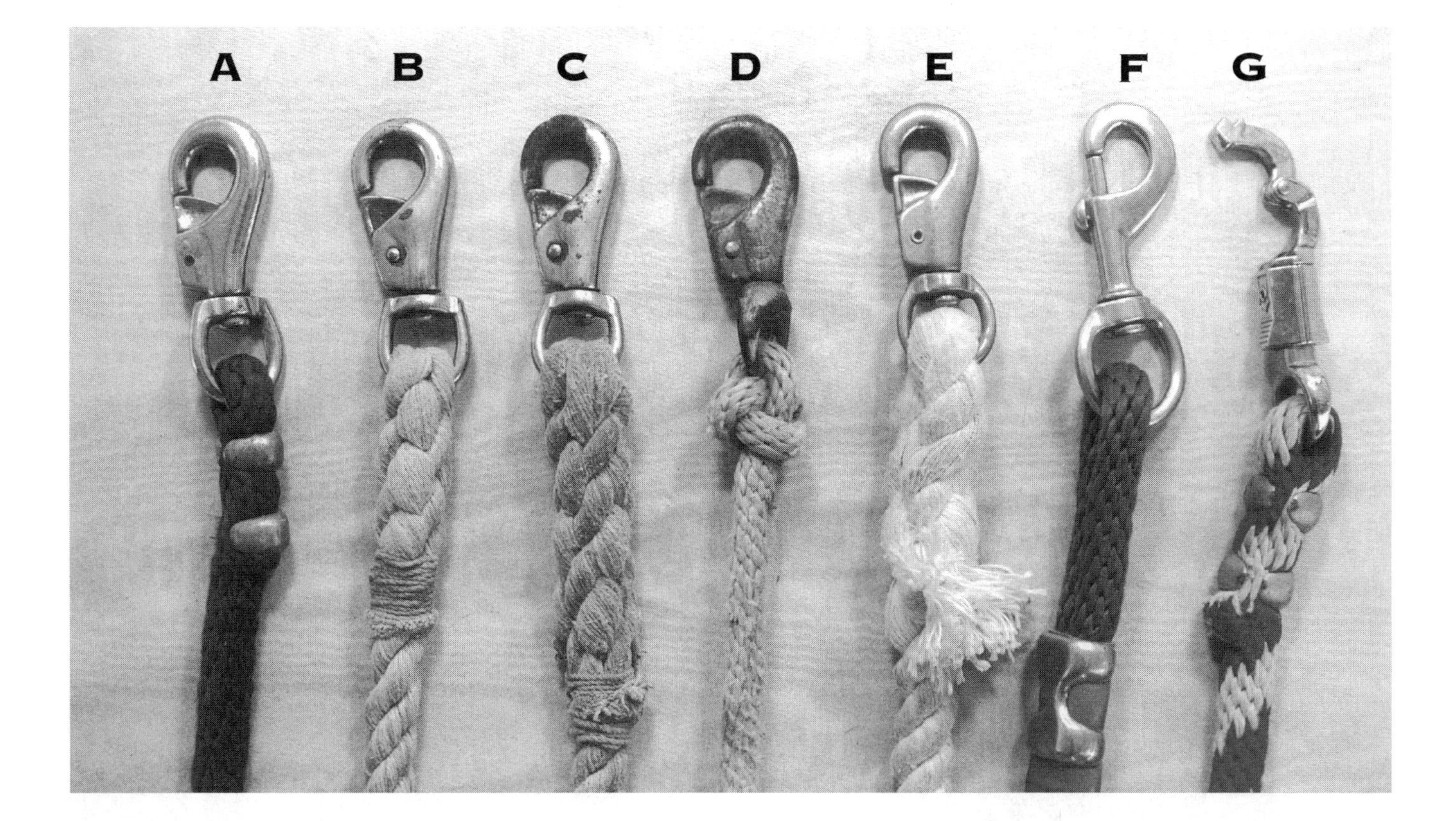

Snaps and Attachments

A **¾" soft nylon rope with rope clamp and new, heavy-duty, chrome-plated-steel bull snap.** This would be an ideal rope for tying a horse outdoors, but it would be more difficult to grip than cotton. Bull snaps are difficult to unsnap when there is pressure on them, such as when a horse is pulling or has fallen and is hanging from a rope.

B **⅝" cotton rope with back splice and whipping and 6-year-old heavy-duty chrome-plated steel snap.** This bull snap is just starting to show rust. This is the most ideal lead rope for in-hand work.

C **⅝" cotton rope with back splice and whipping.** This back splice and whipping is starting to come undone. The 8-year-old heavy-duty chrome-plated-steel bull snap is showing serious signs of rust. This snap might be unsafe to use.

D **11⁄16" hard nylon rope with knot and an 8-year-old zinc-plated bull snap.** The snap is fine but the knot could come untied, and the rope is difficult to hold and tie.

E **¾" cotton rope with back splice but no whipping and a 2-year-old solid brass bull snap.** Note fraying of end of back splice: If whipping were added to this rope, it would be the ideal lead rope in the group for a large hand.

F **⅝" soft nylon rope with leather and brass rope clamp and solid brass spring snap.** This is an easier snap to operate than a bull snap, but this style of snap is more likely to break if a horse pulls.

G **¾" soft nylon rope with zinc rope clamp and chrome-plated-steel panic snap.** A panic snap can be released even when there is pressure on it, such as when a horse is pulling or has fallen while tied. Be careful when leading with a rope having this type of snap so that you don't accidentally release the horse by grabbing onto the collar of the snap.

Haltering and Unhaltering

Approaching and Catching

Keep safety in mind when approaching and catching a horse or foal. Before you approach a horse to catch him, be sure you have the halter unbuckled and the lead rope ready. Always approach a horse confidently. And don't try to sneak up on him; this may spook him.

If you are working with a foal, show that you are confident in how you approach and handle him.

▲
Holding a Foal
To hold a foal, put your left arm in front of his chest to stop his movement. Don't exert pressure as long as the foal stands still. Lean over the foal slightly so the foal feels reassuring body-to-body contact. Keep the foal from backing up by putting your right arm behind his rump. As the foal relaxes, release the pressure from your arms until eventually the foal is standing still on his own with your arms lightly encircling him.

Approaching from the Front

1 When approaching a horse from the front, don't walk directly toward his head or you might cause him to turn away from you. Gaze indirectly at his shoulder instead of peering intently at his head. If you have strong eye contact with your horse at this moment, it could cause him to turn away. Walk toward his shoulder in a nonthreatening manner and say, "Whoa."

2 Reach toward the horse's shoulder with your hand as you get closer. Note that this horse has not moved even one leg from the previous photo, but he has turned his head toward the handler to be caught.

3 Reach your left hand under the horse's neck and transfer the end of the lead rope to your right hand. Note that the horse still has not moved.

4 Bring the end of the rope over to the near side of the horse and make a loop around his neck to hold him while you halter.

Approaching from the Rear

1 When approaching a horse from the rear, speak to him and stay off to the side so he can see you. If you walk up right behind him, you could be in his blind spot and he might spook or even kick if you surprise him. As you get closer to him, turn toward his shoulder.

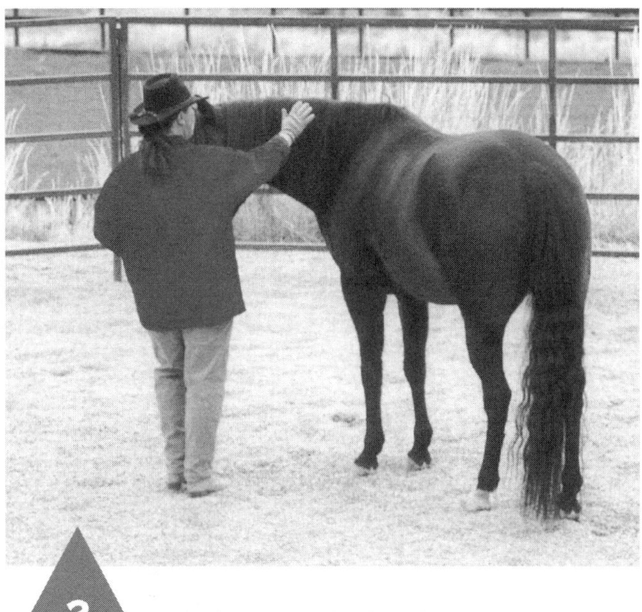

2 Reach your right hand out to touch him on the neck or shoulder.

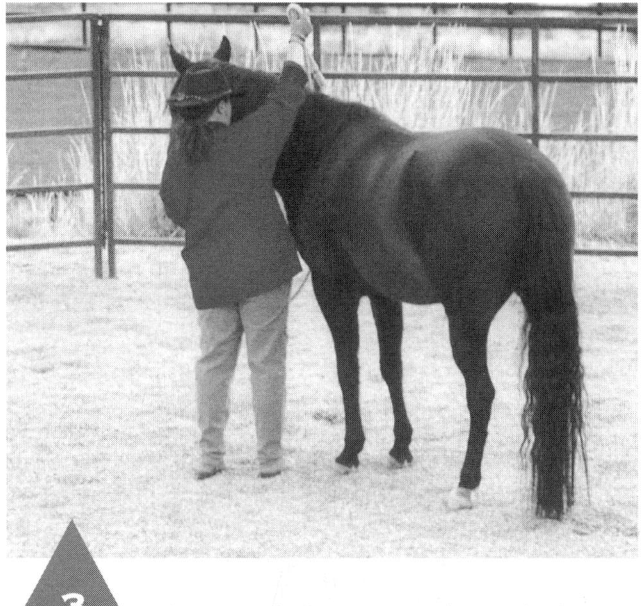

3 Then, as before, pass the end of your lead rope under the horse's neck to your right hand and make a loop around his neck to hold him while you halter.

Putting on a Halter

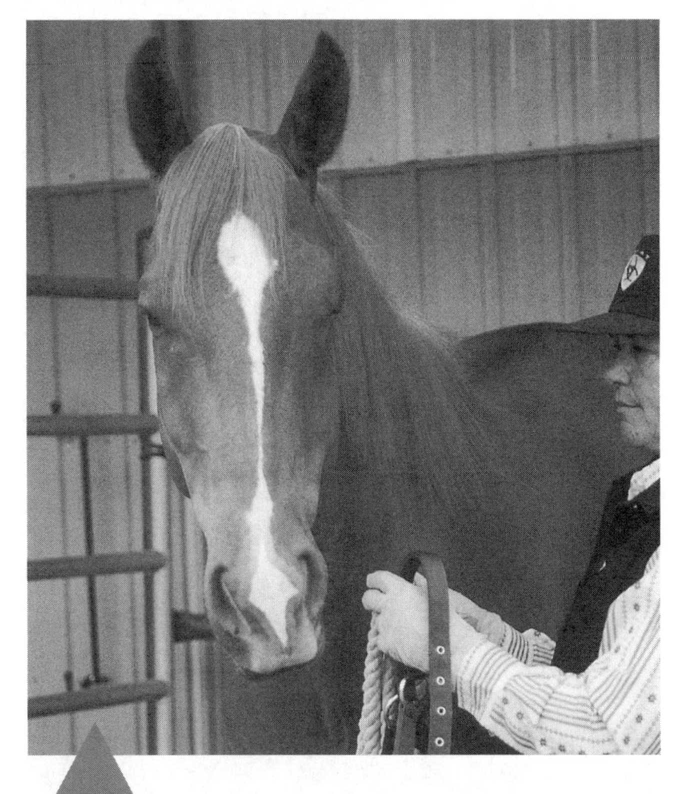

1 When you approach your horse to halter him, have the halter unbuckled and the lead rope ready to put around his neck.

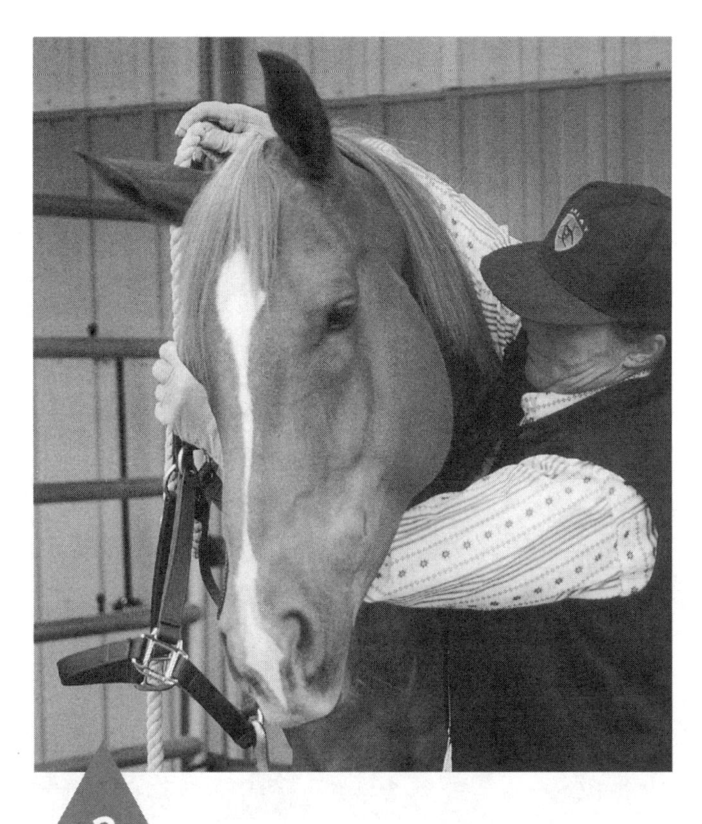

2 Pass the end of the lead rope under your horse's neck to your right hand.

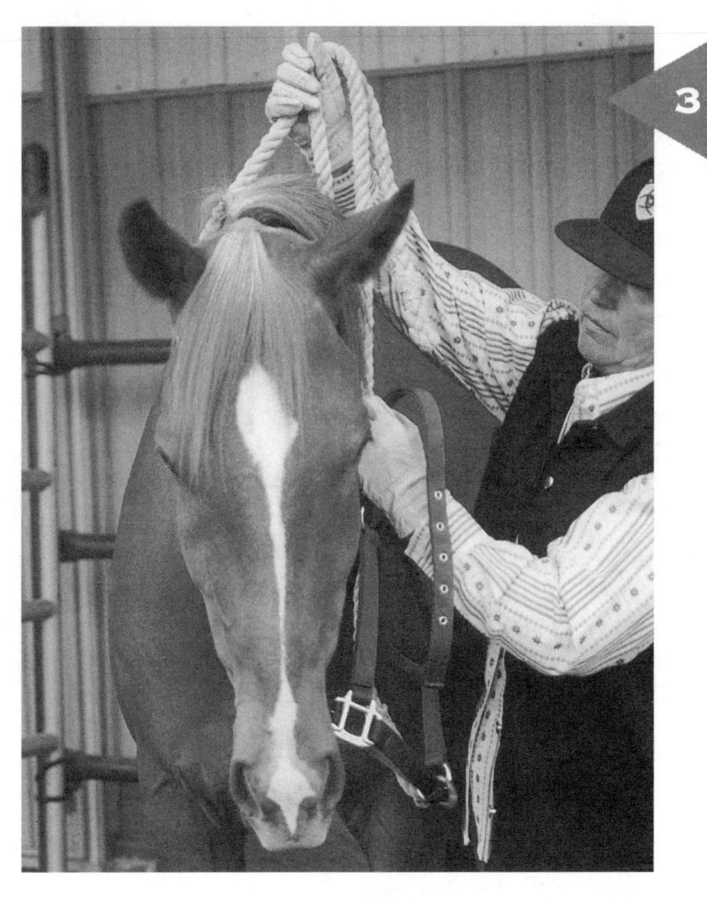

3 Make a loop around the horse's neck so that you can hold him while you are haltering.

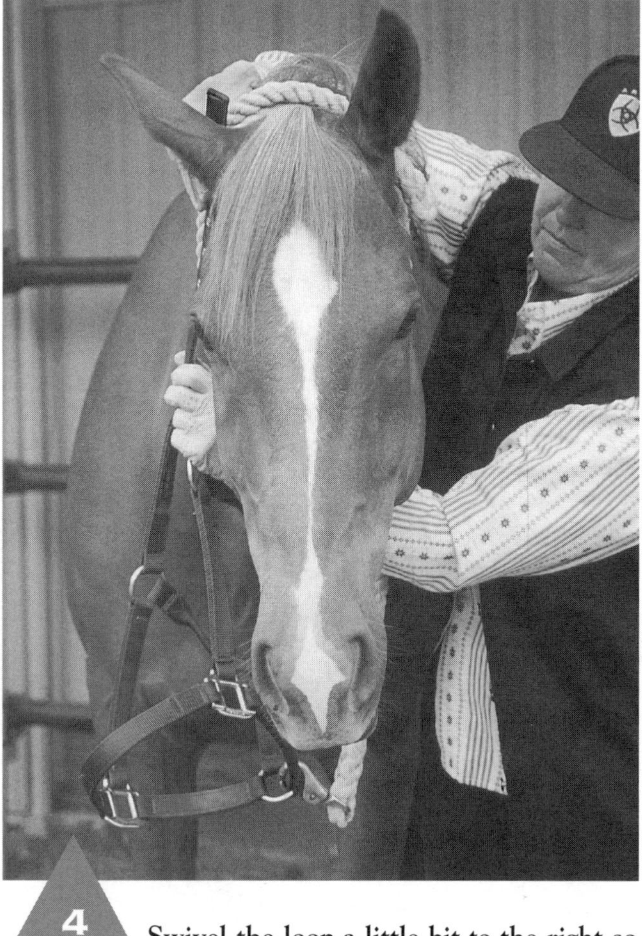

4 Swivel the loop a little bit to the right so you can now pass the halter strap to your right hand, the same way you did with the lead rope.

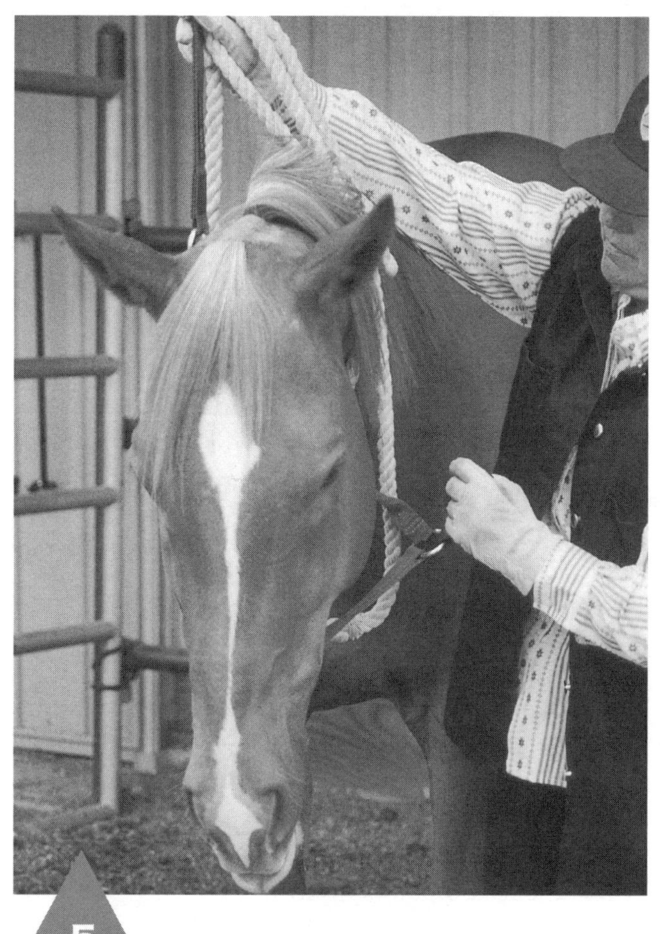

5 Now your right hand is holding the loop and the halter strap and your left hand is holding the halter buckle.

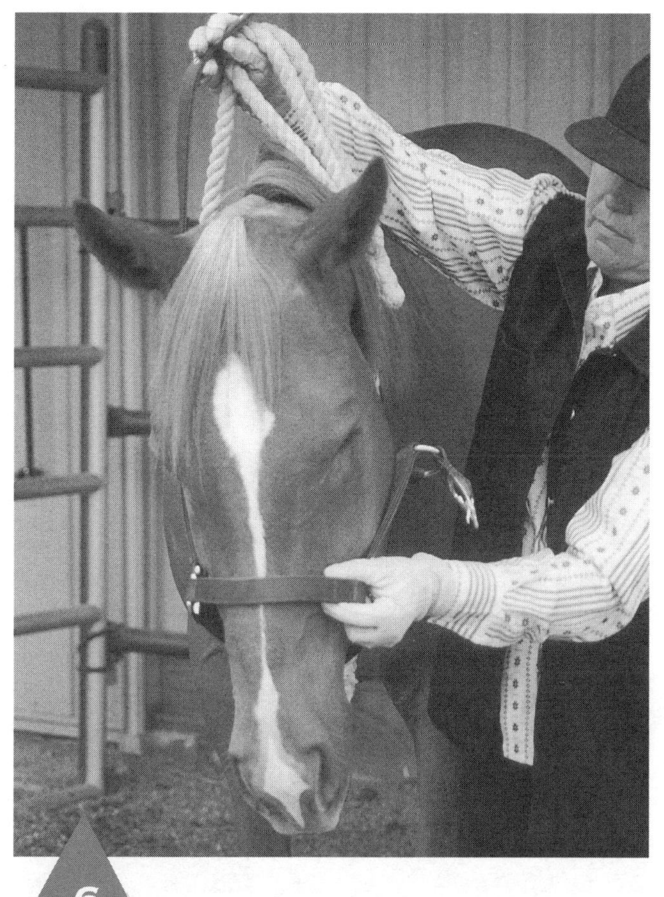

6 Let go of the buckle momentarily and use your left hand to place the noseband of the halter in position.

7 Then take the buckle with your left hand and buckle the halter, still retaining control of the horse with the loop around his neck.

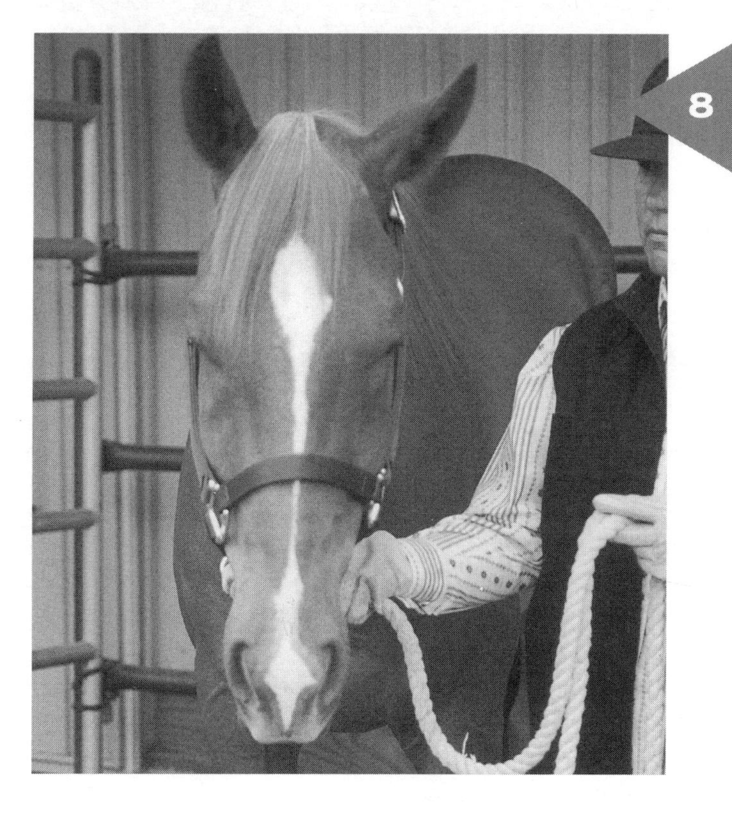

8 Remove the loop from around the horse's neck, and you are ready to lead.

Taking off a Halter

Learn to use the lead rope to help you retain control of your horse as you unhalter and release him.

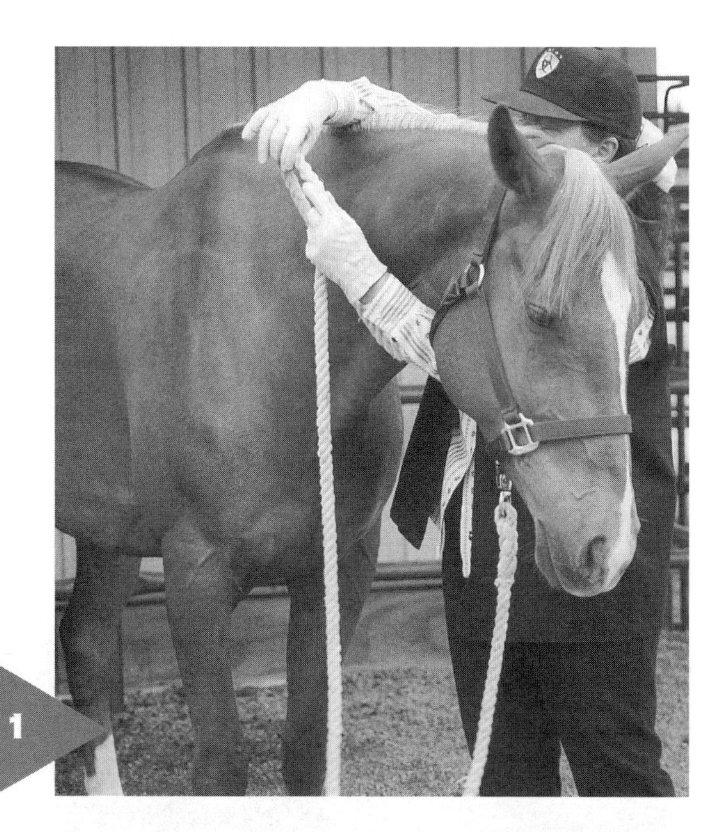

1 To unhalter your horse, first reach under his neck and pass the end of the lead rope to your right hand.

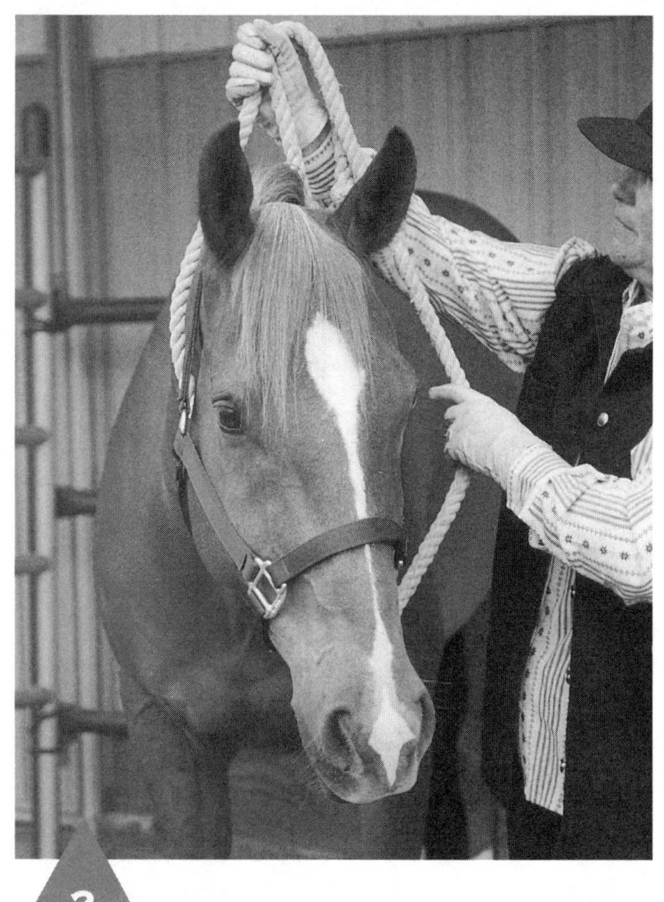

2 Make a loop around his neck with the lead rope.

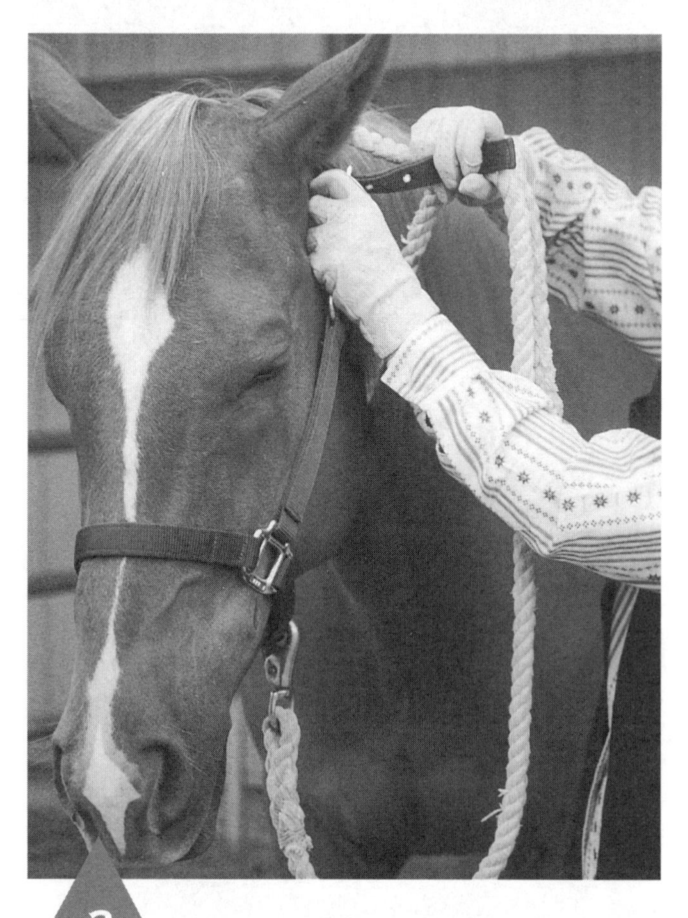

3 Unbuckle the halter while holding onto the loop you have made with the lead rope.

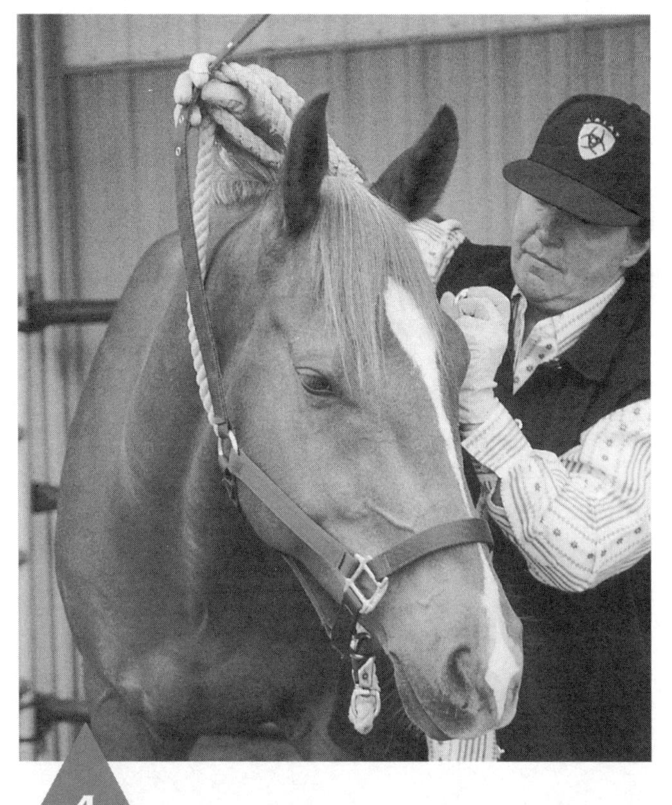

4 Let the halter drop off the end of the horse's nose while you still retain control with the loop.

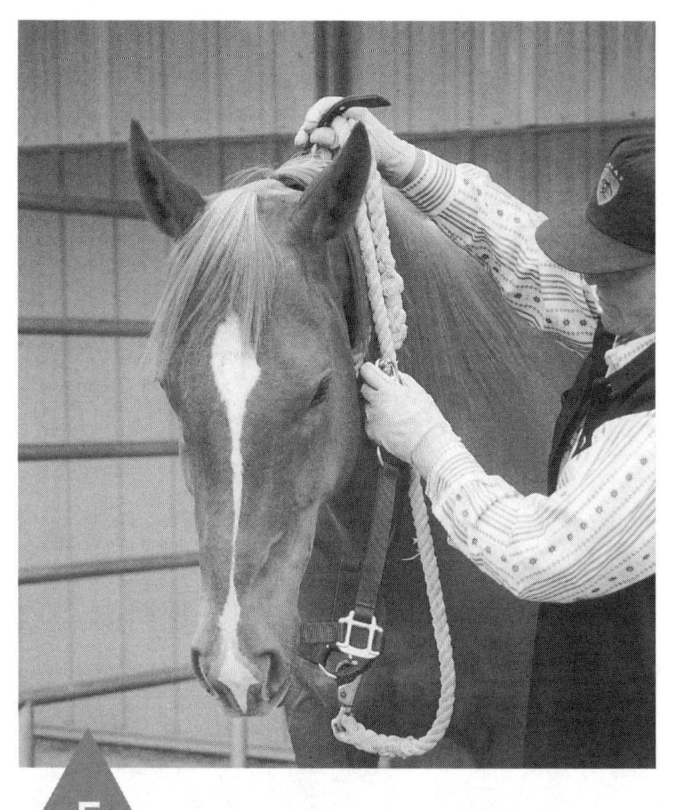

5 Hold the horse momentarily with the halter off and the loop in place so he doesn't develop the habit of moving away as soon as he feels the halter slip off his face.

6 Take the loop from around his neck, but hold onto him for a second before you release him.

TURNING LOOSE

When you turn a horse loose, it is best to do it out in the open away from gates and fences so you both have plenty of room to move away from each other safely.

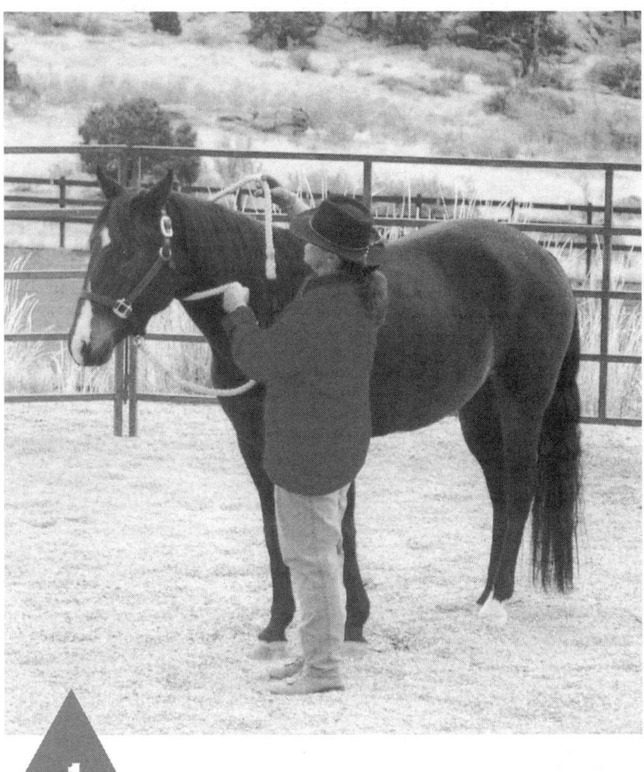

1 Stop the horse relatively square, tell him "Whoa," and then make a loop around his neck with the lead rope.

2 Hold him with the loop while you drop the halter off his face.

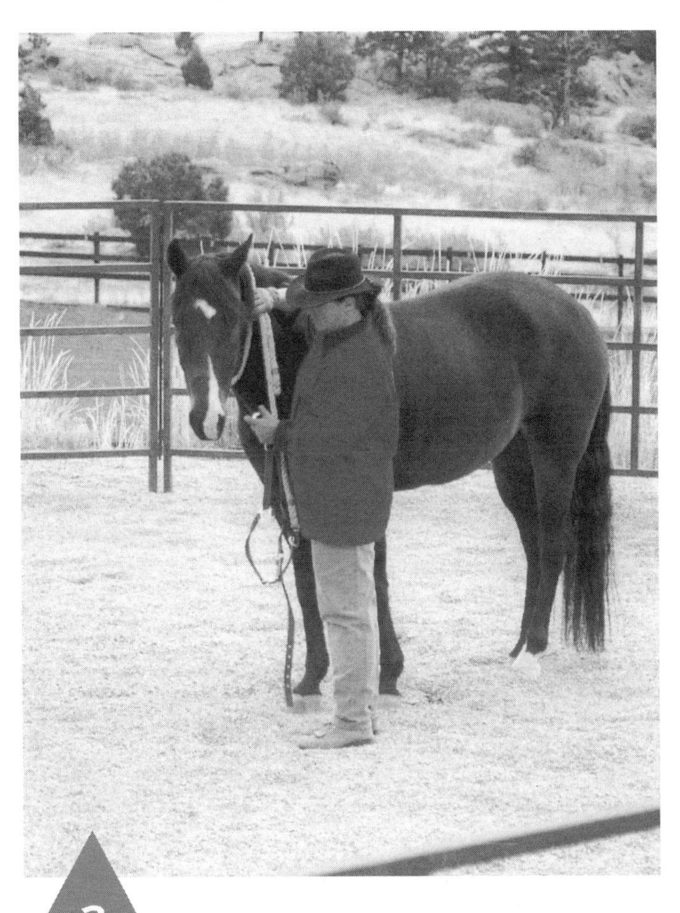

3 Hold the horse momentarily with the rope around his neck but don't let the halter dangle too low; you or the horse could become tangled in it or step on it.

4 Be the one to walk away from the horse; don't let the horse be the one to move away from you before you're ready.

Handling the Haltered Horse

Leading

▲
WEAR GLOVES FOR IN-HAND WORK
When doing in-hand work, always wear gloves to protect your hands from rope burns. Don't roll the excess lead rope into a circular coil because your hand could become trapped inside the coil if the horse suddenly bolts. Instead, fold the excess lead rope in a figure 8 and hold it with your fingers outside the 8. Then, if the horse suddenly lurches forward, the rope will slide cleanly out of your hand one loop at a time. You can carry a 30-inch whip in your left hand along with the excess lead rope. Hold the lead rope about 6 inches from the snap with your right hand as shown here. You can let the rope slide through your fingers and take up slack by letting your fingers walk up the rope.

▲
HAND POSITION ON ROPE
If you are working with a young horse or one who constantly charges forward, hold the rope with your right hand like this instead. This gives you more "whoa power."

▲
LEADING STRAIGHT AND TURNING LEFT
If you are leading with a bridle, hold the reins separated by your index finger so you can operate the reins independently when you turn. When you are leading straight ahead, hold the reins as shown in this photo. To turn left when leading from the near side, continue holding the left rein between your thumb and index finger but open the rest of your hand so the right rein can slide through your fingers as you bring your hand toward your body to turn.

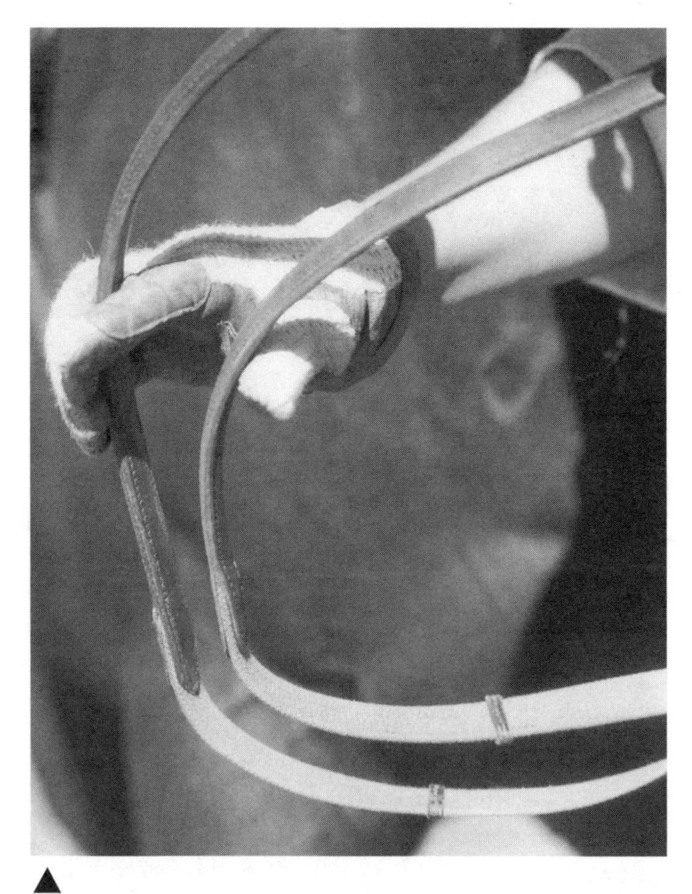

▲
TURNING RIGHT
To turn right when leading from the near side, open the space between your thumb and index finger so the left rein can be free but keep hold of the right rein between your index finger and your middle finger as you push your hand under the horse's neck to the right.

▲
Using Your Elbow
If your horse starts to crowd you when you are leading (this is often a sign of insecurity), tell him to keep his distance by bumping the side of his neck with your elbow.

▲
Leading at a Walk
When leading at the walk, move forward energetically as if you have somewhere to go. Look ahead. Maintain a position at the midpoint of the horse's neck, between the horse's head and shoulder. If the horse lags behind, give him a light tap with the whip on his hindquarters. Never try to hurry him forward by pulling on the lead rope; pressure on the halter means slow down or stop. If the horse charges ahead, give a couple of light tugs on the lead rope. If he is really charging ahead, turn him to the right, into the fence, to slow him down.

◀ **Leading at a Trot**
When leading at a trot, maintain your position at the midpoint of the horse's neck. Be sure you are holding the lead rope 6 to 10 inches from the snap so the horse has room to trot up. Ask the horse to begin trotting at the same time you do by whatever means you customarily use: "Ta-rot!" or a click-click sound with your tongue against your teeth, or a light tap with the whip on his hindquarters. This is the first step of trot with the whip in a perfect position to use if necessary, but this horse trotted forward immediately.

Backing Up

There will be times when you want to get a horse to back up. And once a horse learns the command, responding to it becomes automatic. The following steps apply to teaching the horse to back up initially. It also works with a horse who has already learned to back up on command.

1 To get a horse to back up, stop him, then turn around and face his shoulder. Change the lead rope and whip to your left hand. Place the tips of the fingers of your right hand on the point of his shoulder and apply light pressure as you say "BAAAACK." Don't use halter pressure to back your horse; this usually results in him raising his head and hollowing his back. This photo shows the first step of a back where the horse starts out by moving one hind leg backward. Next he will pick up his left front leg and his right hind leg at the same time and move them backward.

2 The horse then will pick up the other diagonal pair of legs and move them backward in unison. As long as you want the horse to continue backing, keep the light pressure on the point of his shoulder. Backing him along a fence will help keep him straight. If his hindquarters start to swing off the straight line, straighten his body with a slight adjustment of his head position with the lead rope. If his hindquarters start moving off to the right, move his head slightly to the right to bring his hindquarters back on track.

TURNING

When you are handling a haltered horse, you will need to communicate to the horse to turn right or to turn left, or to turn on the forehand or hindquarters. The body language and signals you use are an important handling skill and are well worth learning and practicing.

▲
TURNING RIGHT
To turn your horse to the right when you are leading from the customary near side, use body language and proper signals to help your horse know what you want. Push your right hand under the horse's neck to the right. Note the legwork of the handler and the corresponding response from the horse.

▲
PRACTICE ON BOTH SIDES
You should practice everything with your horse from both sides so he doesn't become "one sided." Here is an example of leading from the off side, turning to the left. Note the body language and footwork of the handler. This yearling, like most horses, is stiffer when first worked from the off side. With time, he will perform a smooth arcing turn from either side.

▲
TURNING ON THE FOREHAND
To turn on the forehand from the near side, hindquarters moving to the right, tip your horse's nose to the left, slightly toward you, to weight the left front leg. Then cue him on the side with very light intermittent pressure to cause him to pick up his left hind leg and cross it over and in front of his right hind leg. His next step will be to uncross his right hind leg and step to the right with it. The hind legs will continue crossing and uncrossing as long as you continue the cues. Meanwhile, the left front leg stays in the same spot and the right front leg walks a small circle around it. (See **Leading through a Gate** for other views of this maneuver.)

▲
TURNING ON THE HINDQUARTERS
To turn your horse on his hindquarters to the right with you on the near side, first stop your horse. Then with your right hand on the lead rope, push to the right and backward at the same time to settle the horse's weight on the right hind leg. Walk around the horse in a small circle. The left front leg will cross over and in front of the right front leg, then the right front leg will uncross. The right hind leg will stay relatively stationary while the left hind leg walks a small circle around it. (See p. 27 for another view of this maneuver.)

Leading through a Gate

When taking a horse through a gate, it is best to do so in a controlled, progressive manner. Otherwise, a horse tends to rush through narrow openings and swing quickly around, creating the potential for injury to himself as well as to his handler.

To Pass through a Gate That Swings toward You

1 Stop your horse and then back him up as you begin opening the gate.

2 When the gate is fully open, let go of the gate and walk your horse straight through.

3 Stop your horse in a relatively square position and prepare for a turn on the hindquarters to the right.

4 Perform a turn on the hindquarters to the right, causing your horse to pivot on his right hind leg.

5 Reach for the gate and close it, taking care not to bump your horse's nose, which should have ended up very close to the gate.

To Pass through a Gate That Opens Away from You

1 Stop your horse at the gate and unlatch it.

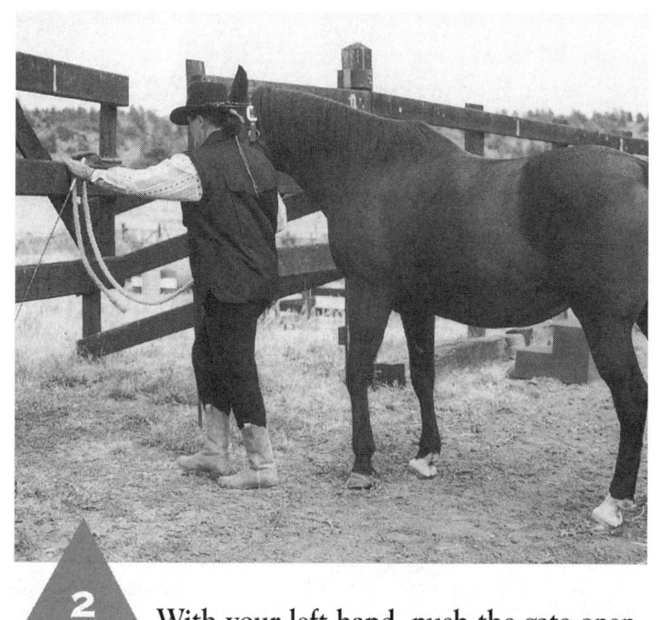

2 With your left hand, push the gate open.

3 Walk your horse through the gate.

4 Stop and perform a 180° turn on the forehand to swing your horse's hindquarters to the right.

5 Halt your horse when he is facing the opposite direction from where you started and close the gate, taking care not to bump his nose.

6 Lead your horse a few steps ahead to latch the gate.

Tying

Safe Tying Practices

Whenever you tie a horse, there is the potential for danger if he pulls, falls, or becomes trapped. That's why it is essential that you use safe tying practices.

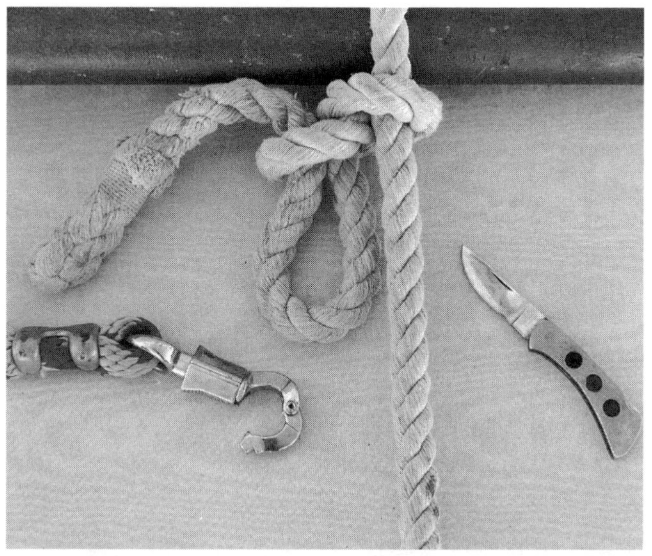

▲
PANIC SNAP AND QUICK-RELEASE KNOT
Although with horses that are accustomed to being tied in most cases you can use bull snaps or trigger snaps on the rope, consider using a panic snap (left) if you are tying a horse who tends to be nervous when tied. Tie all horses with a quick-release knot so if you need to, you can instantly pull the end of the rope and release the horse. Keep a sharp pocketknife close at hand when you tie a horse; if all else fails, you can cut the rope if the horse is trapped. Practice using the knife beforehand on a thick, tight rope so you know how to wield it without injuring yourself or your horse.

▲
A SECURE TIE RING
When tying to a wall, be sure the tie ring is strong and well attached to the wall. This tie ring is bolted to the wall with six 4-inch lag bolts.

▲
Tie to a Strong Rail

Always tie a horse at the level of his withers or a bit higher to a post or a well-constructed rail, such as this metal pipe rail. Wooden fence rails are usually not suitable for tying; a pulling horse could easily detach them and panic when the rail hits him or chases after him.

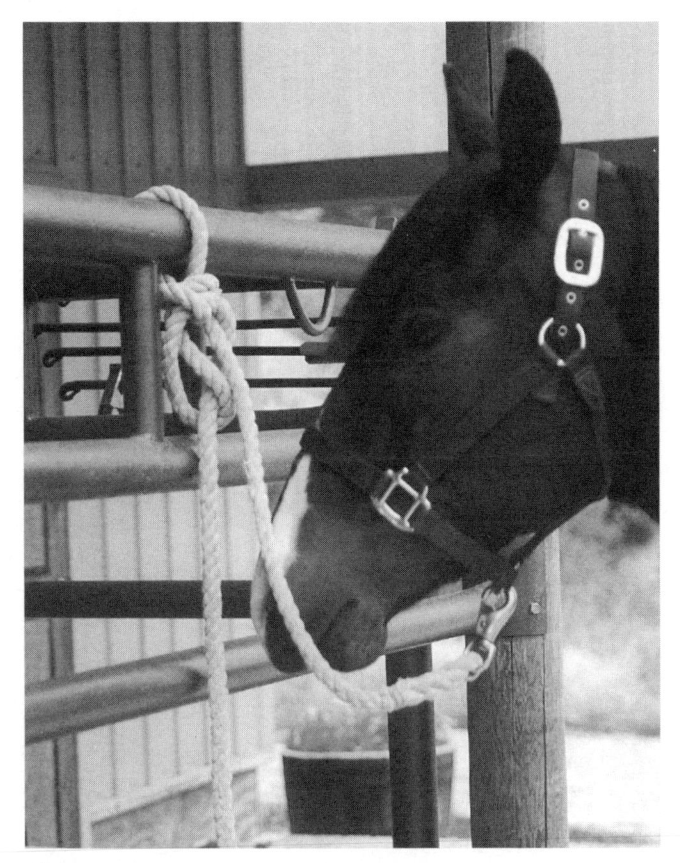

▲
Use a Quick-Release Knot

This horse is tied at a proper height and at a proper length with a quick-release knot. (The tail of the rope has been dropped through the loop of the knot so the horse cannot release himself.)

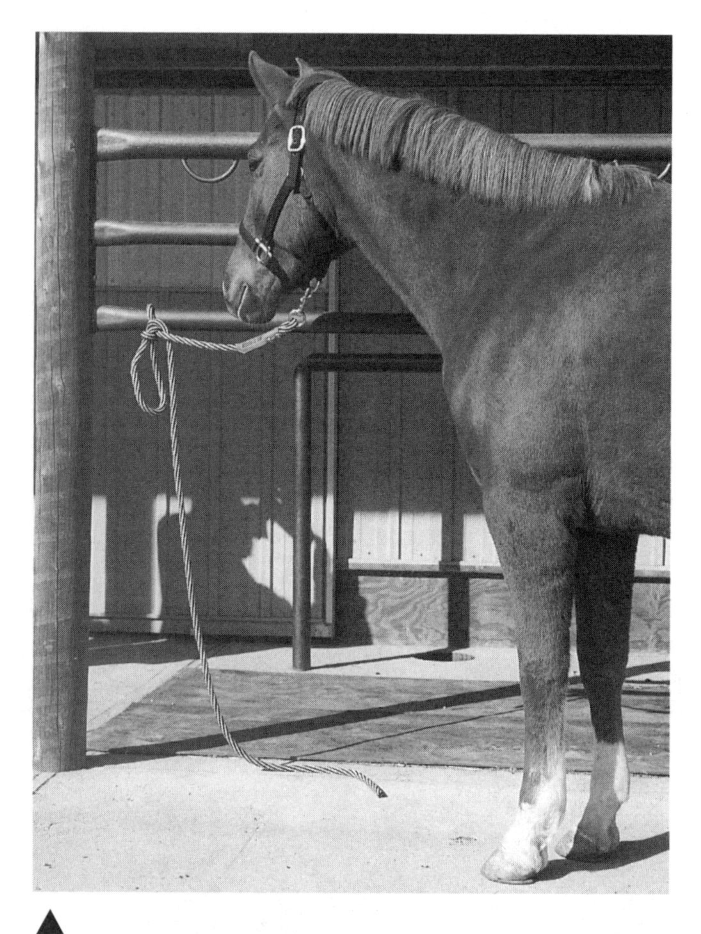

▲
PROBLEM: TIED TOO LOW
This horse is tied to a rail that is below the level of his withers. If he wanted to pull, he could get some very good leverage with his front feet. He should be tied to the top rail, which is a few inches higher than his withers. The length of the lead rope is fine but the long tail dangling on the ground is just inviting trouble. This horse could release himself by pulling the tail of the rope because the tail has not been dropped through the loop of the knot.

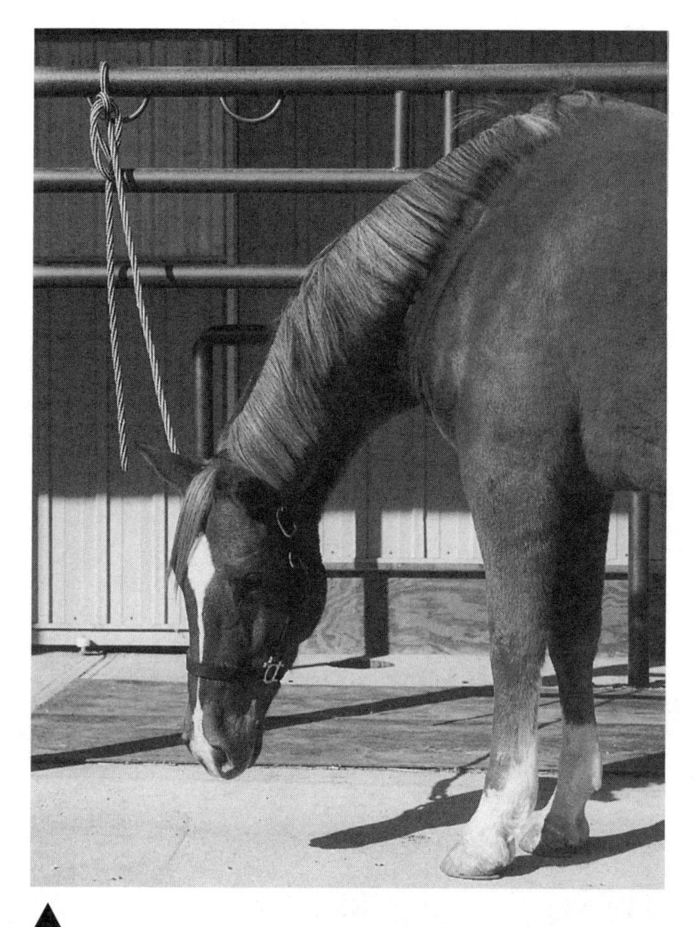

▲
PROBLEM: TIED TOO LONG
Although this horse is tied at the proper height, the rope is much too long, which allows him to lower his head to the ground, move around too much, and possibly get his front leg over the rope.

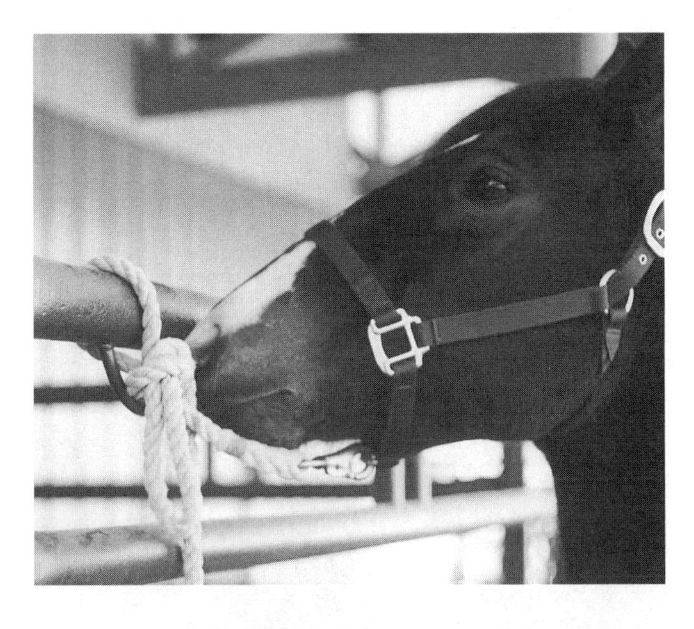

◀ PROBLEM: TIED TOO SHORT

Although this horse is tied high enough, the lead rope is too short, causing the horse to hold his head and neck in an uncomfortable, cramped position. Tying this short could cause even a well-trained horse to pull.

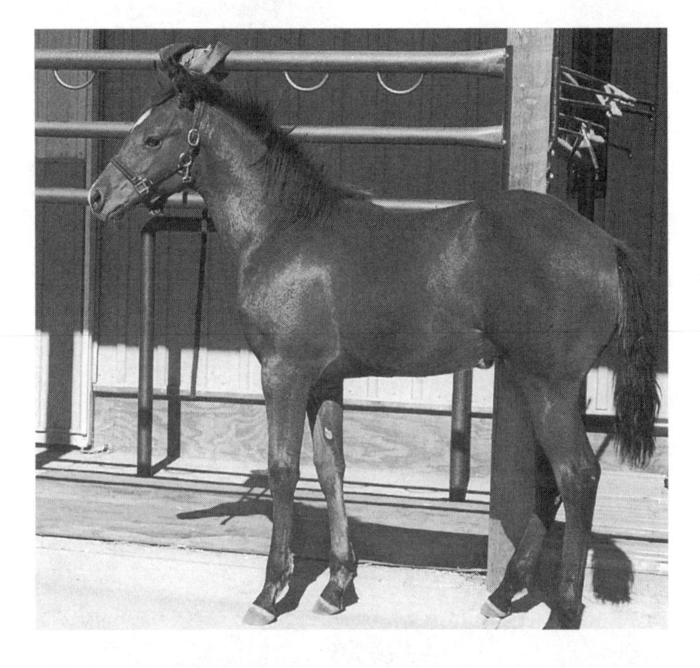

◀ TYING A FOAL

When tying a foal for the first few times, do not tie him "hard and fast" to a solid object because if he pulls (which most foals will do at least once or twice) he could injure his poll or neck. Instead, tie the foal's lead rope to a car or truck tire inner tube, which will act like a fat rubber band and provide some "give" when the foal pulls. This 5-month-old foal is tied properly in a safe area and has learned to stand patiently.

◀ TYING TO AN INNER TUBE

This is a close-up showing the use of the inner tube. In this instance the foal's rope is about the right length to allow him to hold his head at a natural level. If the rope were shorter, it would cause the foal to hold his head up in a unnaturally high, fatiguing position.

To Use the Inner Tube with a Tie Ring

When you will be tying a young horse or a horse not accustomed to being tied, tie him to an inner tube attached to a ring securely mounted on the wall.

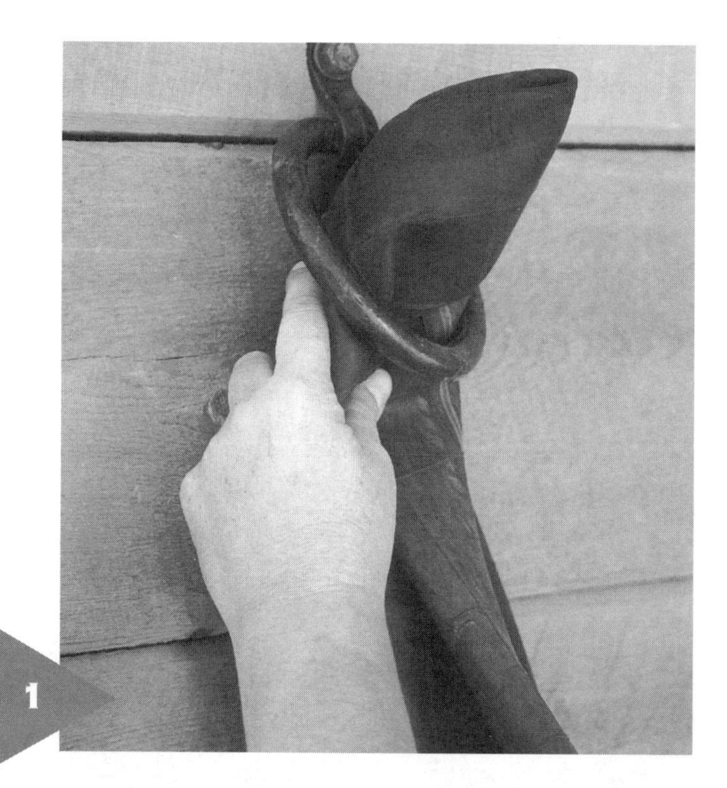

1 Place the inner tube through the ring from back to front, until half the inner tube is hanging in front of the ring and half is hanging behind the ring.

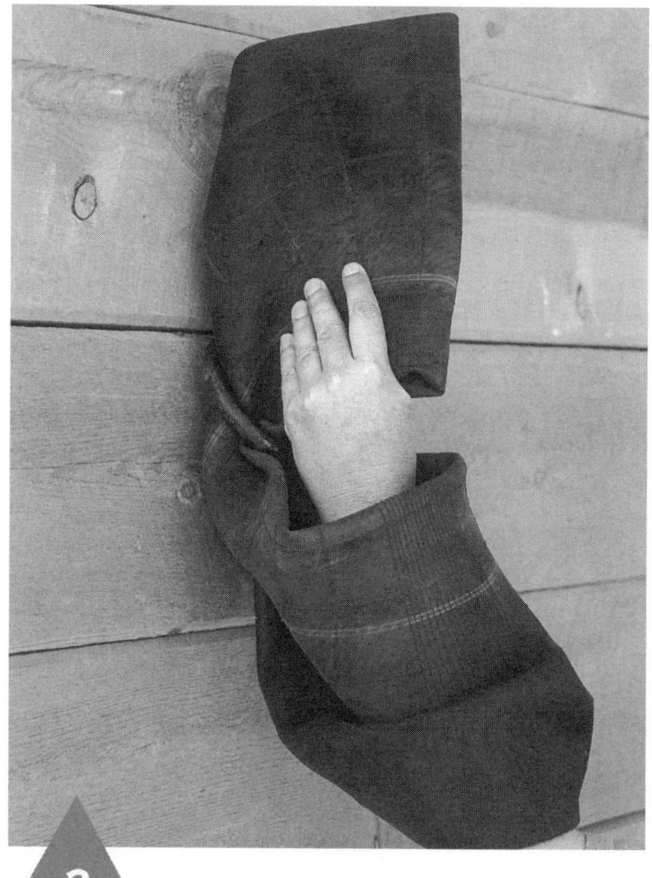

2 Reach through the bottom half of the inner tube and grab the top half.

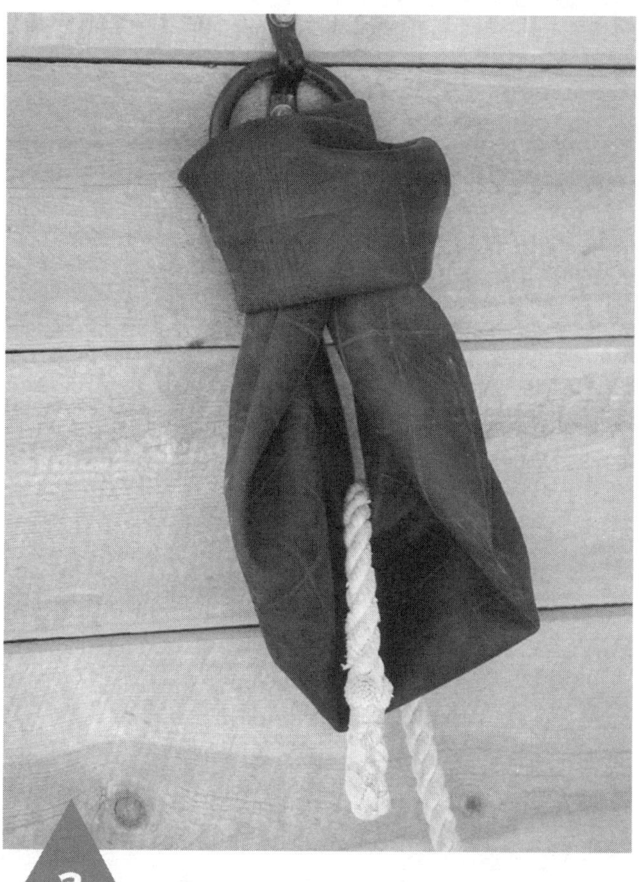

3 Pull the top half through the hole in the bottom half until it looks like this. Tie the lead rope to the lower loop of the inner tube.

Tying a Horse in Cross Ties

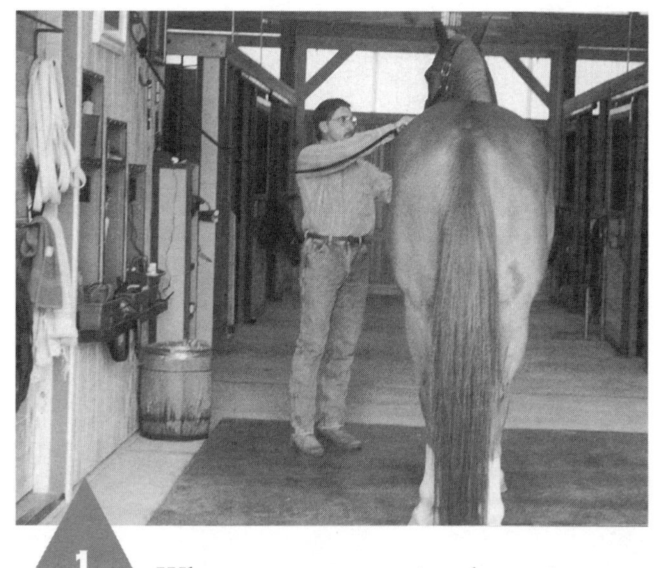

1 When you want to tie a horse in cross ties across a barn aisle, stop the horse relatively straight with his head at the cross ties. Turn and face the rear of the horse, with your left hand holding the lead rope. With your right hand, grab the near cross tie. Fasten the snap to the near cheek ring on the side of the halter, not the halter ring at the throat of the halter.

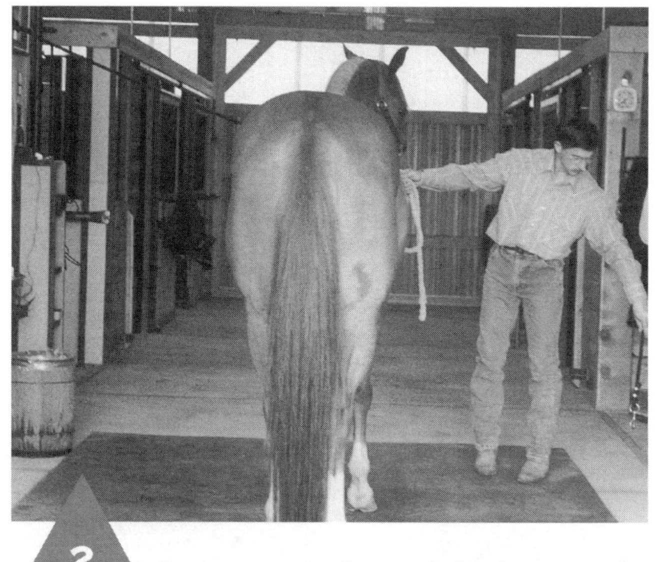

2 Step across in front of the horse to the off side, maintaining his position with the halter rope with your right hand, and reach for the far cross tie with your left hand.

3 Fasten the snap to the off cheek ring of the halter. Remove the lead rope.

A CROSS-TIE RING can be mounted to the wall with ⅜-inch diameter x 2½-inch long lag bolts.

Tying a Quick-Release Knot

1 Run the tail of the rope through the ring. The horse is attached to the portion in your left hand.

2 Hold the two portions of rope together in your left hand.

3 With your right hand, pick up a portion of the tail end of the rope and make a fold (bight) in it like this. Cross the bight over the two ropes you are holding in your left hand and through the loop that has formed.

4 Here the bight has been placed through the loop. Take care not to let your fingers get inside any of the loops; if your horse were to pull at this point, your fingers could get trapped in the loops.

5 Grab the tip of the bight and pull it through the loop.

6 Pull until the U-shaped bight is about 6-inches long and the knot is snug. Then move the tail of the rope from the left side over the right side so it is lying against the wall, out of your way.

7 Grasp the portion of the lead rope your horse is attached to with your left hand and the knot with your right hand. As you pull with your left hand, slide the knot up to the ring with your right hand.

8 Here is the quick-release knot. The horse is attached to the section at the left. The tail end is hanging along the wall and is the portion you would pull to free your horse.

▲
HORSEPROOF FOR EXTRA SECURITY
If your horse has learned how to nibble the quick-release knot and free himself, you will have to "horseproof" the knot by dropping the tail of the rope through the loop as shown here.

▲
THE HORSEPROOF VERSION
Here is the horseproof version of the quick-release knot. In order to free your horse, first remove the tail of the knot from the loop. Because of this extra step, this is no longer a real quick-release knot. The horse-proof version of the knot should only be used with horses who have learned how to untie the standard quick-release knot.

Positioning While Tied

To Move a Horse Forward

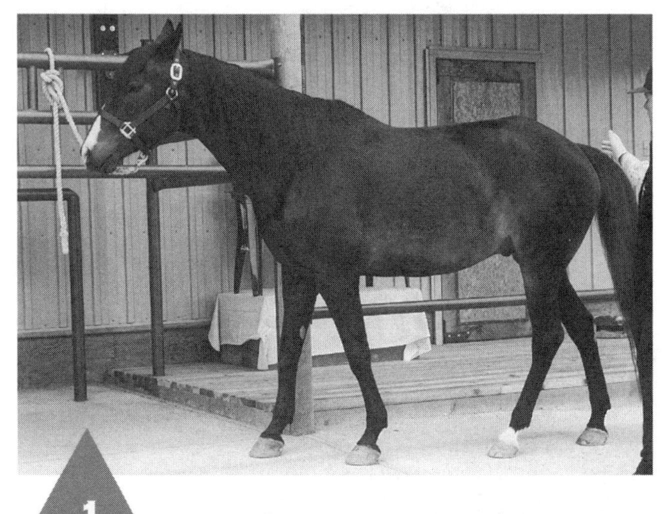

1 When your horse is tied and you want to move him forward a few steps, step behind him, raise your hand behind his rump, and say "Walk up" or make a click-click noise with your tongue against your teeth. If he has had good in-hand lessons, he will move forward with this alone.

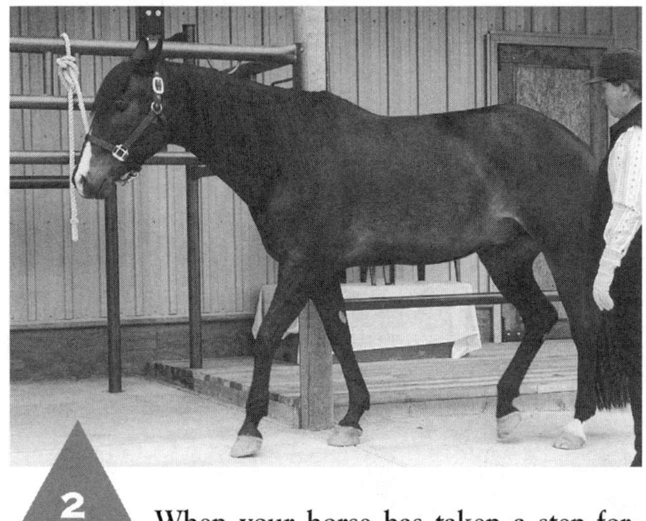

2 When your horse has taken a step forward, immediately lower your hand so he doesn't go too far. You might want to add the word "whoa" if you think he will walk too far ahead.

To Back a Horse Up

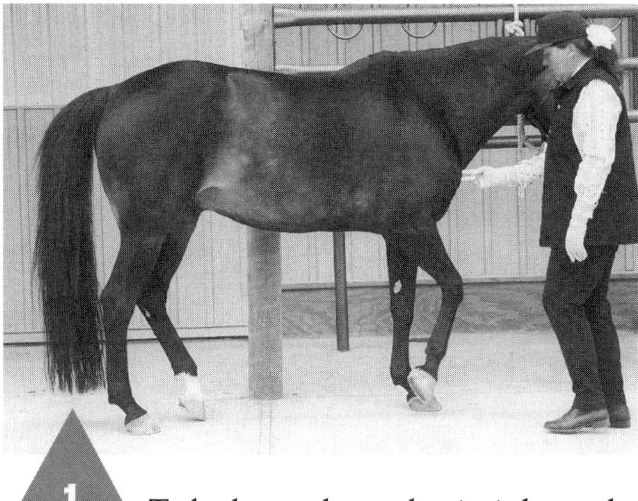

1 To back up a horse that is tied, use the same technique you did during in-hand work. Use your fingertips on his shoulder point while you say "BACK." Here he has lifted the first diagonal pair of legs.

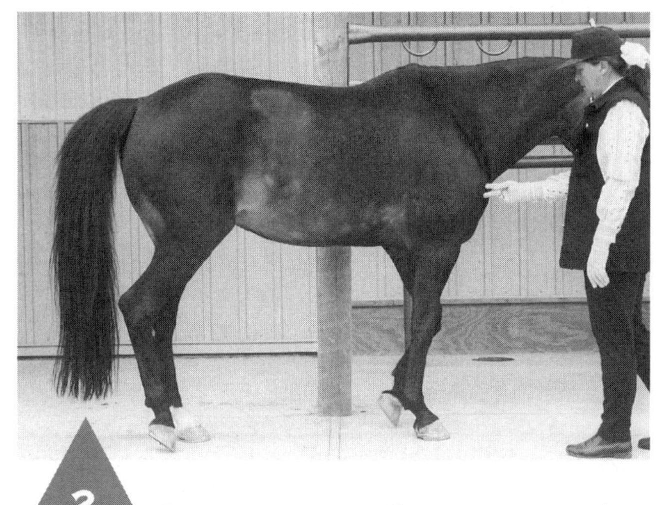

2 Continue cueing as long as you want him to move back, taking care to note the restriction from the lead rope.

To Move a Horse Over

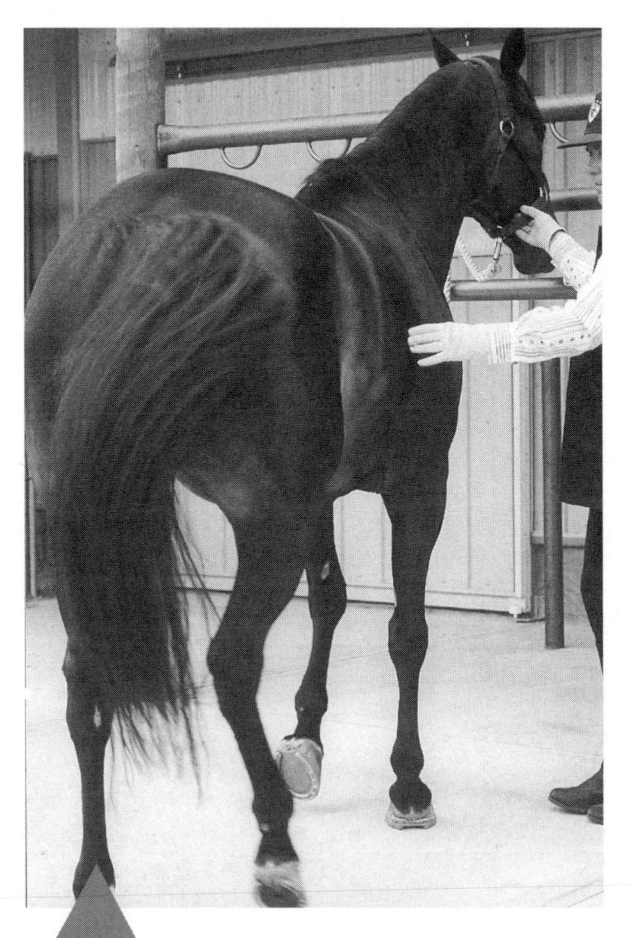

1 To move a horse over while he is tied, use the same technique you used to turn on the forehand during in-hand work. Here the horse is being moved 180° from a position alongside the tie rail to one facing the opposite direction. Slightly tip his head toward you and apply light pressure to his side. His right hind leg is getting ready to cross over the left hind leg and his left front leg is walking a small circle around his right front leg, which is the "pivot" foot in this turn. Note that the pivot foot is facing the "old direction" and is twisting in place, so the horse will have to pick it up to reorient it to the new direction.

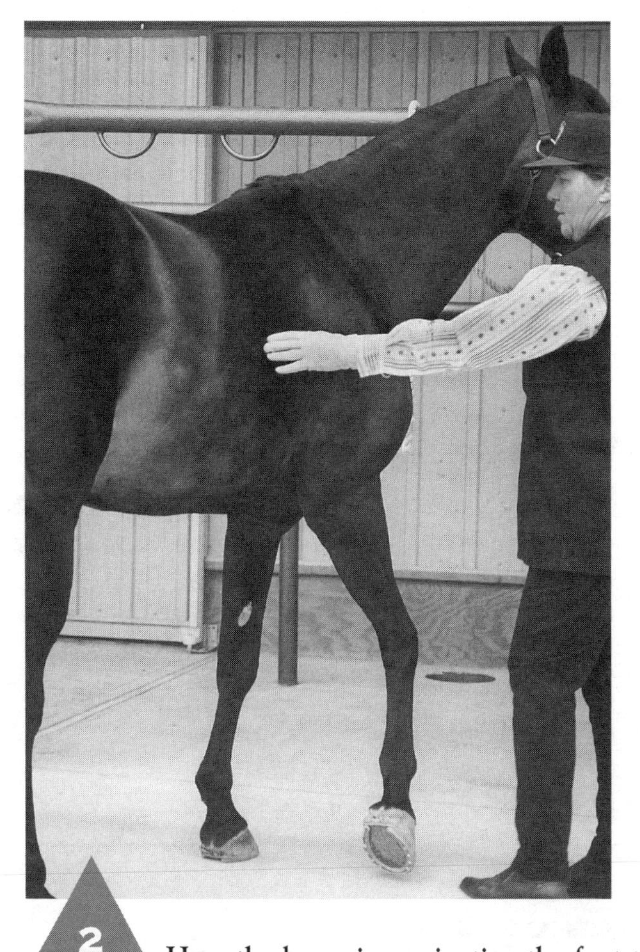

2 Here the horse is reorienting the foot to face the new direction.

Safe Training for Handling

Applying a Halter Chain

Sometimes a chain is used in conjunction with a halter to provide additional control of an unruly animal, an injured animal that must be treated, or for the routine handling of a stallion. Properly applied and used, a chain can prevent a horse and handler from getting hurt. A chain should be used with intermittent pressure and release. A horse should NEVER be tied with a chain.

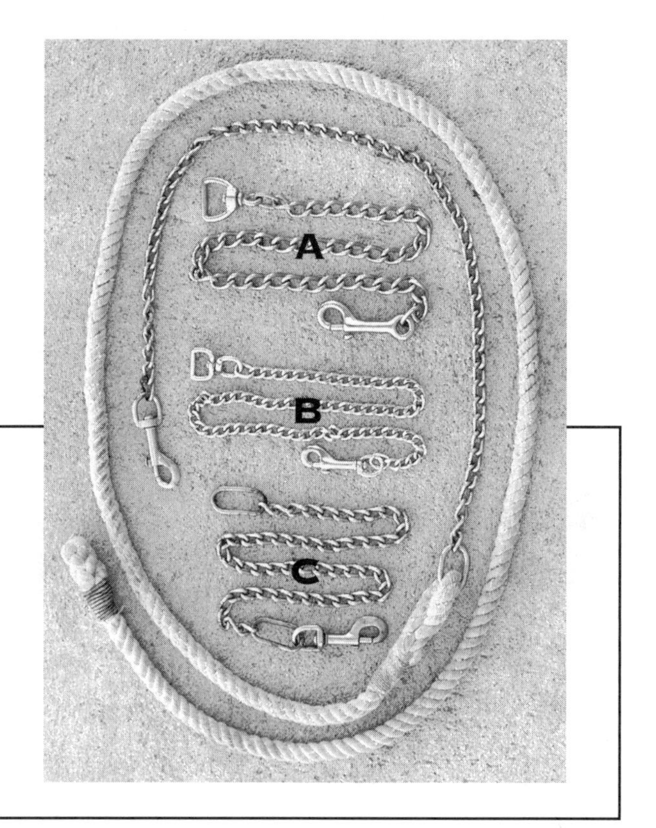

Different Kinds of Halter Chains

A chain can be permanently attached to the lead rope, as with the rope that encircles this group. Most chains are designed to be used separately so they can be used when needed with a variety of lead ropes.

A — a 30" solid brass coarse chain
B — a 30" plated steel fine chain
C — a homemade chain from obstetrics chain, welded links, and snap

Method One

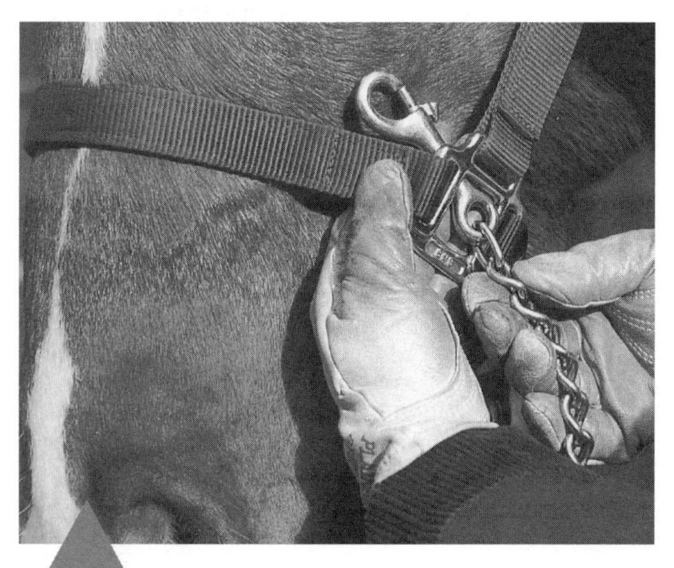

1 There are two basic ways to attach a nose chain. To perform either, hold the horse with halter and lead rope while the chain is attached. In method one, pass the snap up through the near cheek ring from outside to inside.

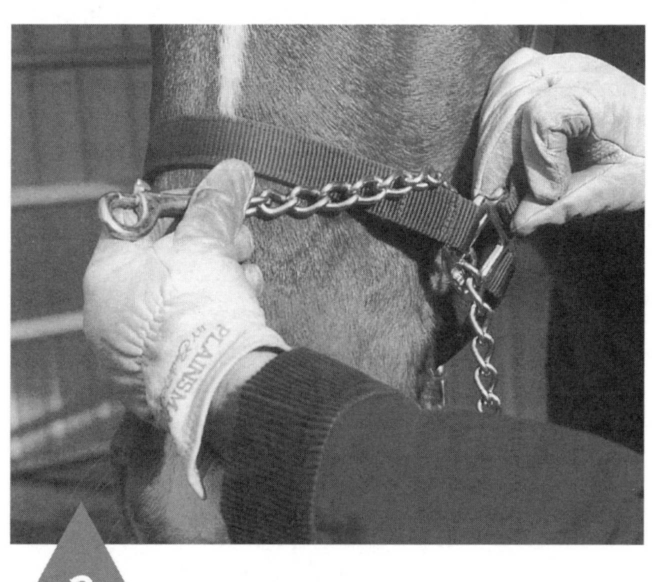

2 Cross the chain over the noseband of the halter. This will prevent the chain from slipping down on the horse's nose when pressure is released.

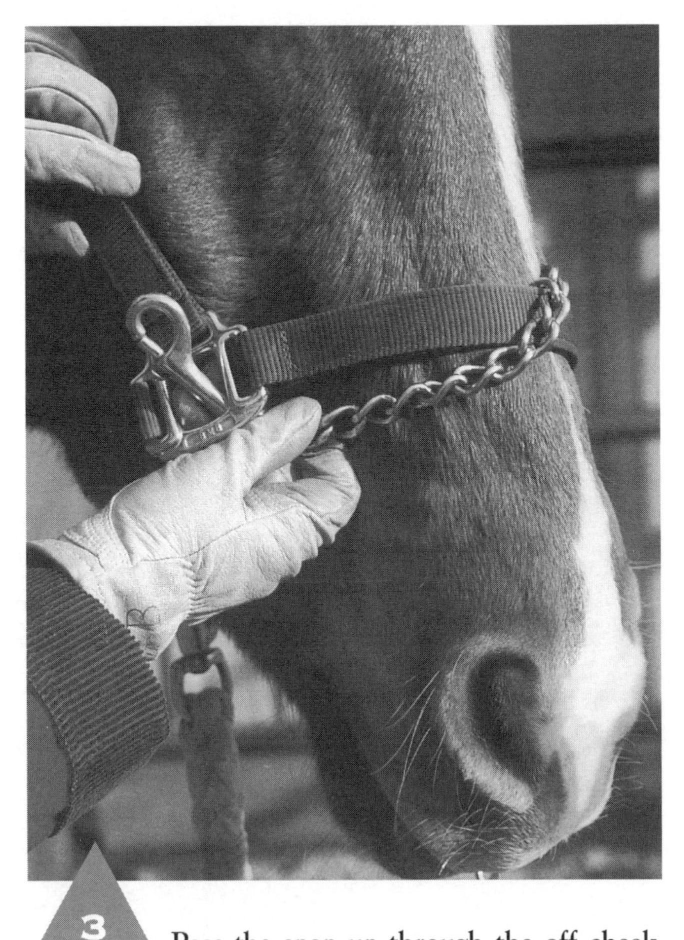

3 Pass the snap up through the off cheek ring from inside to outside.

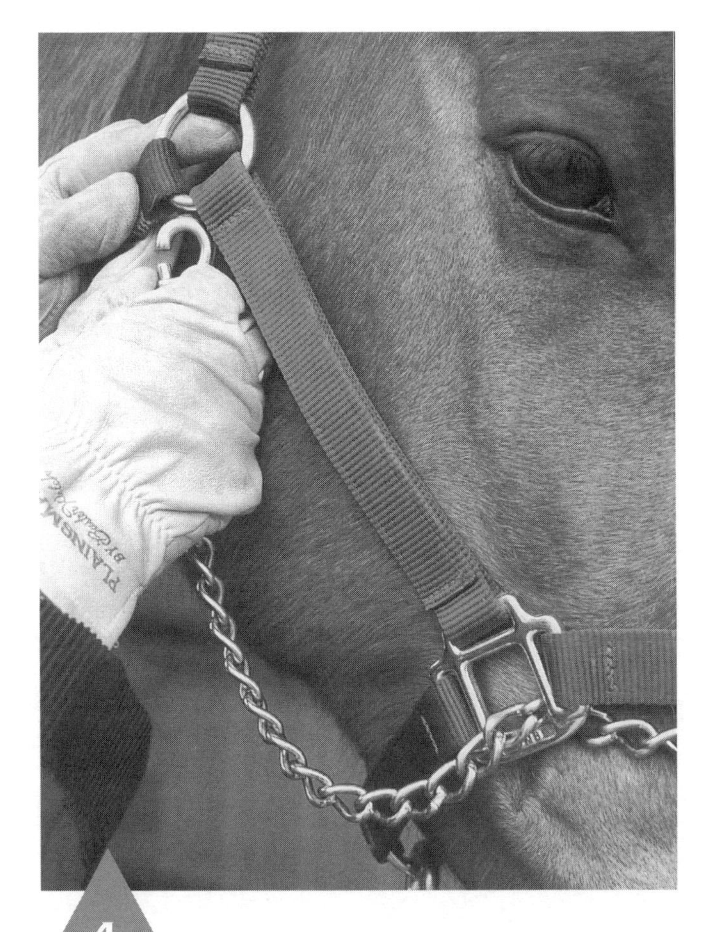

4 Attach the snap to the upper halter ring on the off side.

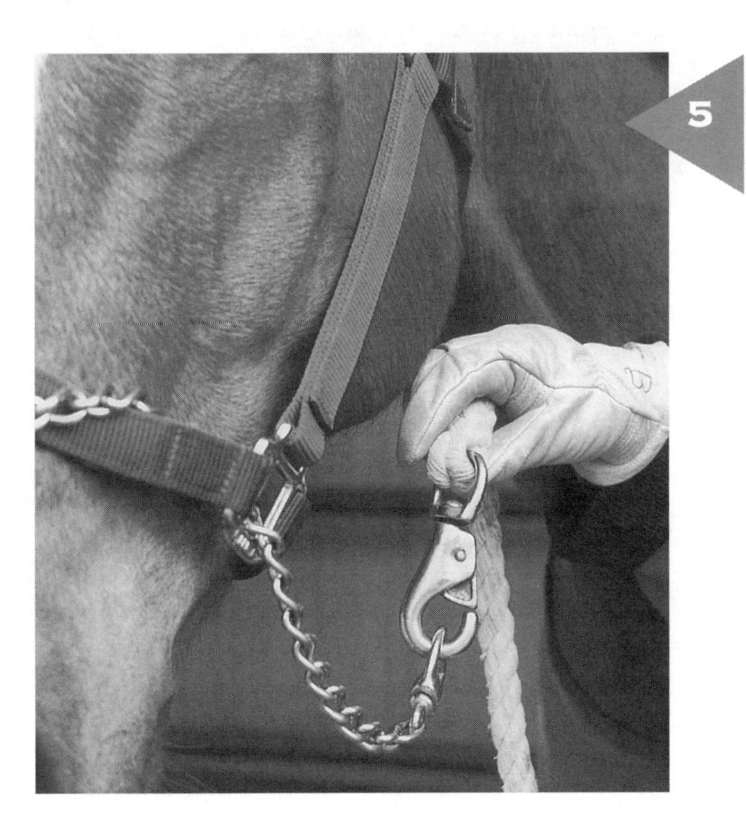

5 Go back to the near side and attach the lead rope snap to the chain.

Method Two

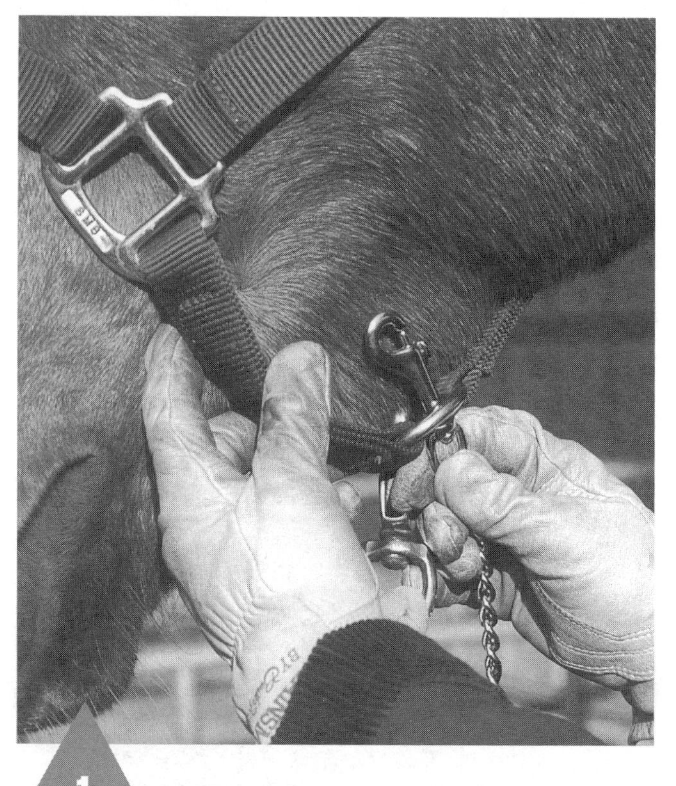

1 Pass the snap up through the throat ring of the halter.

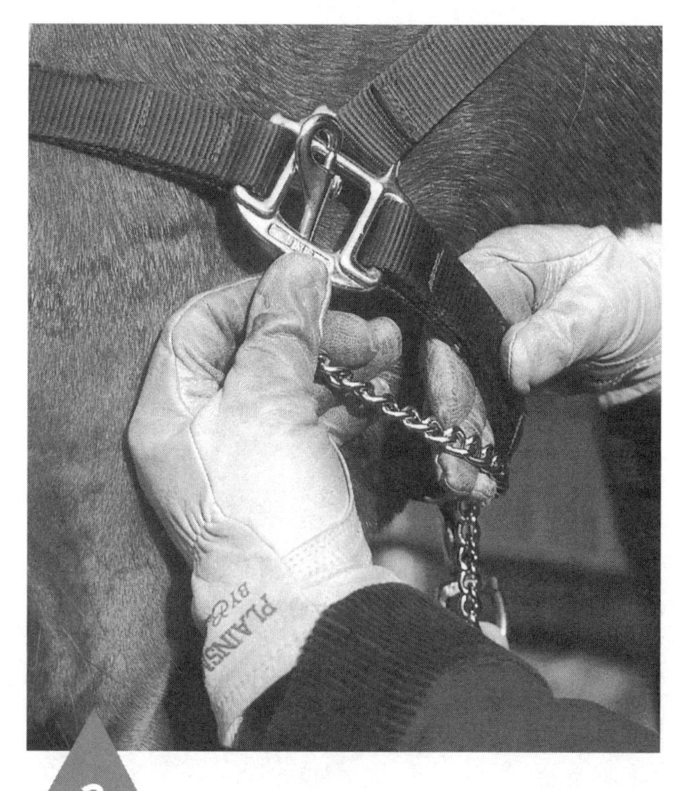

2 Pass the snap from inside to outside up through the near cheek ring.

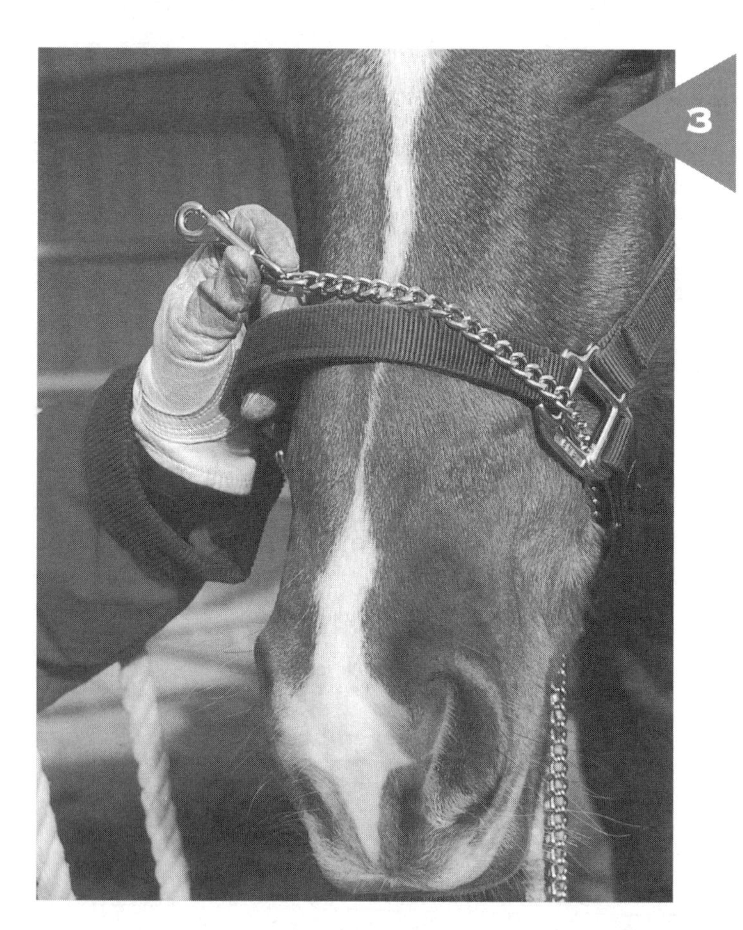

3 Lay the chain over the top of the noseband of the halter. Because you are not taking a wrap over the noseband, be aware that with this method, the chain can slip down the horse's nose.

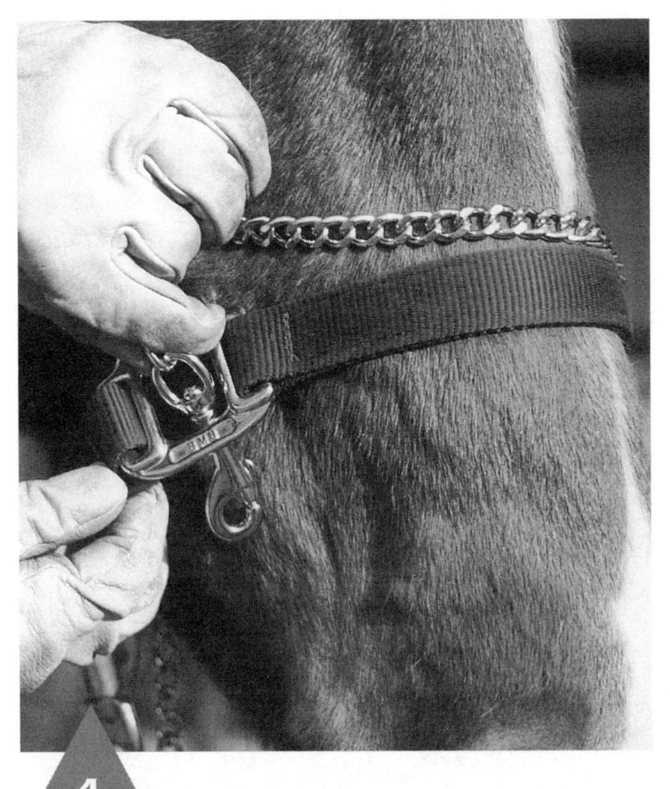

4 Pass the snap down through the off cheek ring from outside to inside.

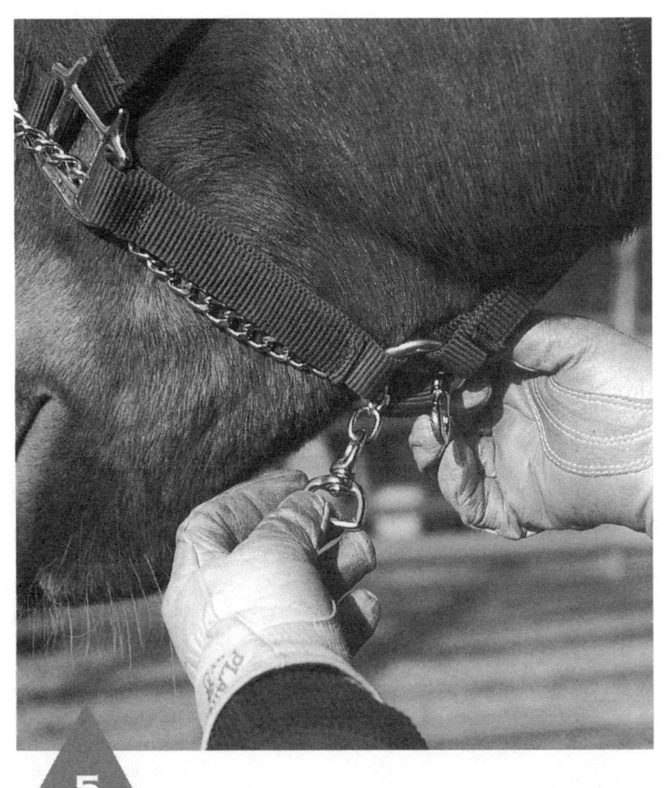

5 Next, pass the snap down through the throat ring.

6 Fasten together the two ends of the chain and snap the lead rope to the chain.

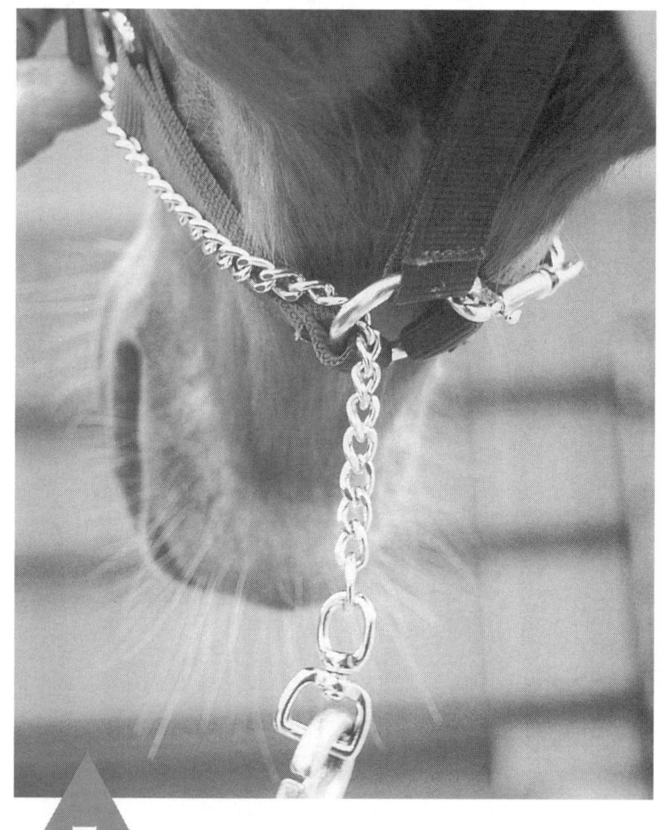

7 If your chain is too short, you have two options. You can snap the nose chain to the throat ring of the halter and snap your lead rope to the other end of the nose chain.

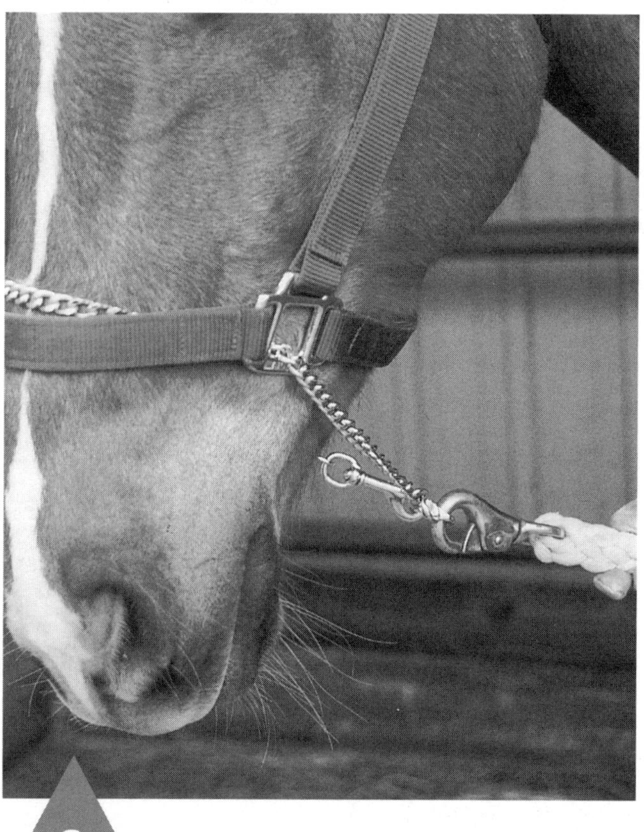

8 Or you can bypass the throat ring altogether and fasten the nose chain to itself and snap the lead rope to it.

Lowering a Horse's Head

Many routine procedures require you to handle your horse's head: dental care, deworming, bridling, clipping, among other tasks. If your horse raises his head out of your reach, you will need to teach him to lower his head. Many horses will learn to lower their heads without the use of a chain. Try the procedure without a chain first. If necessary, use a chain, then eliminate the chain and use the lightest cues possible.

Problem: Horse Raises His Head ▶

If your horse raises his head out of reach, you'll have to teach him to lower it.

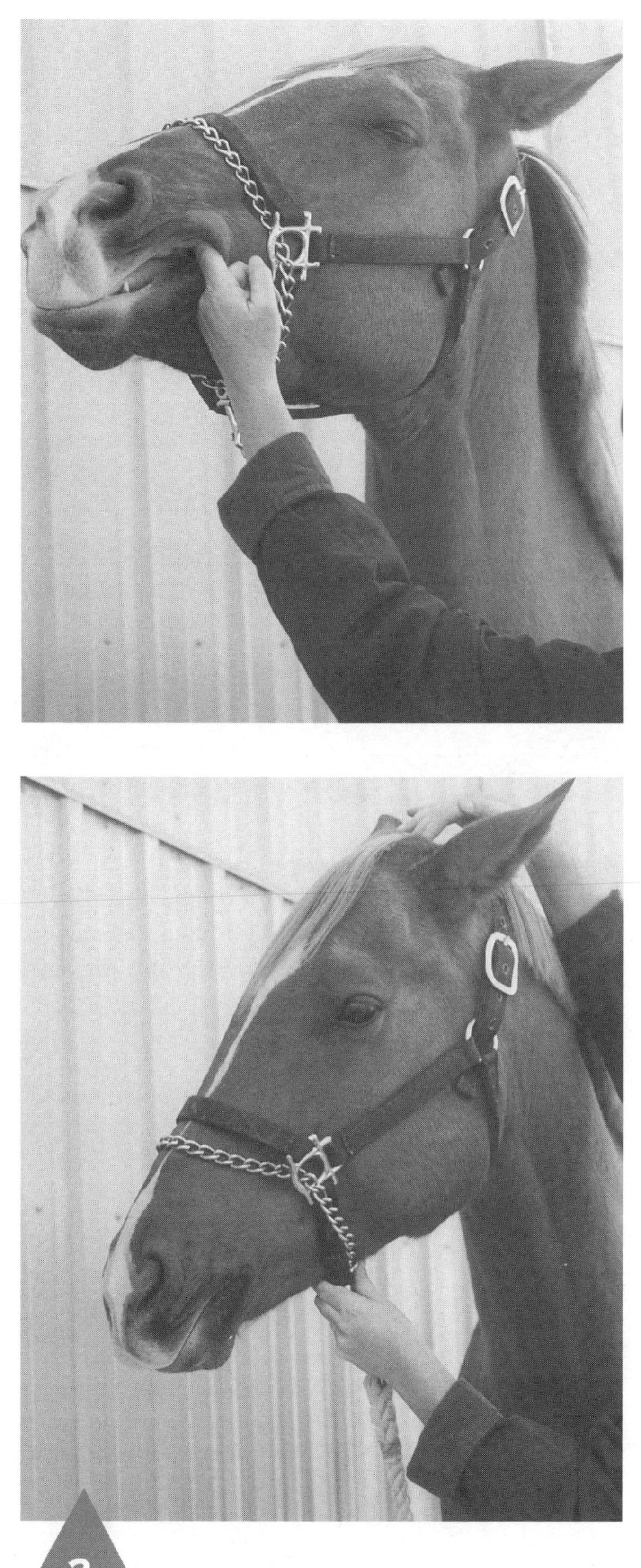

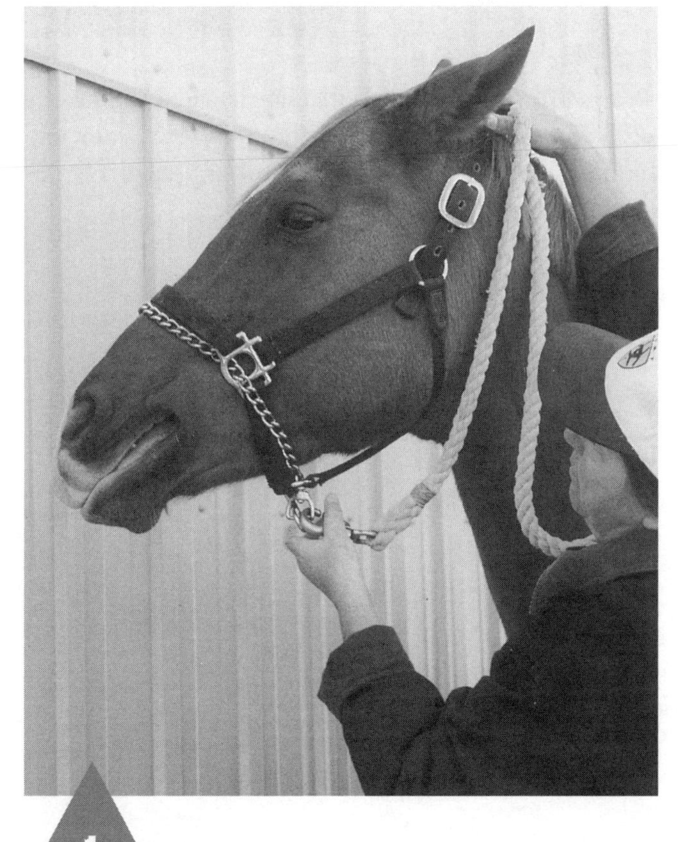

1 With the chain attached, place your fingertips on the horse's poll.

2 Use a command such as "Down" as you apply pressure with the fingers of your right hand.

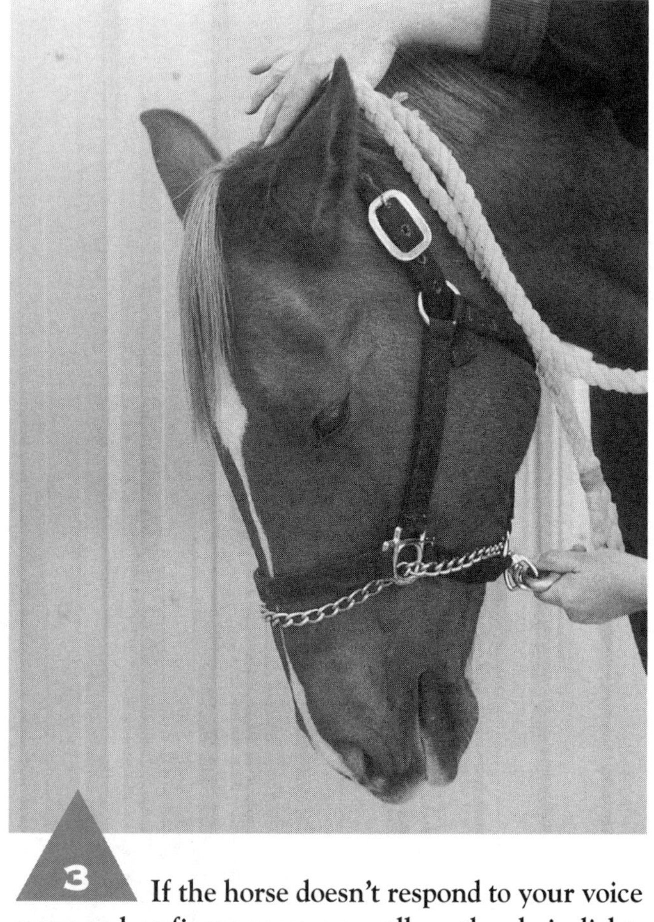

3 If the horse doesn't respond to your voice command or finger pressure, pull on the chain lightly and intermittently with your left hand. As soon as the horse's head starts to lower, stop the pressure with the chain.

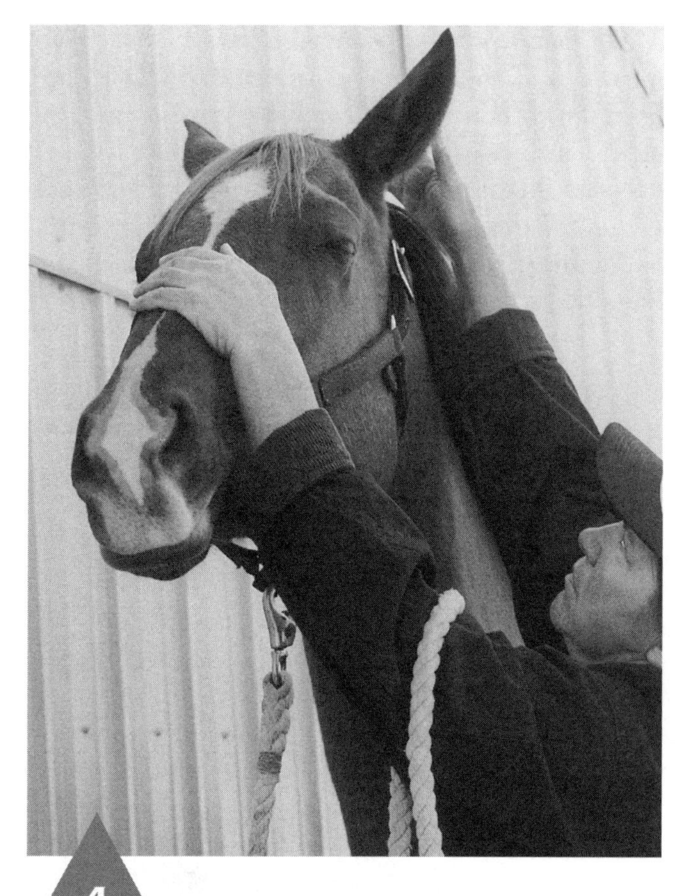

4 When your horse has learned the lesson with the chain, remove the chain and put your left hand over the horse's noseband in place of the chain. Repeat the exercise.

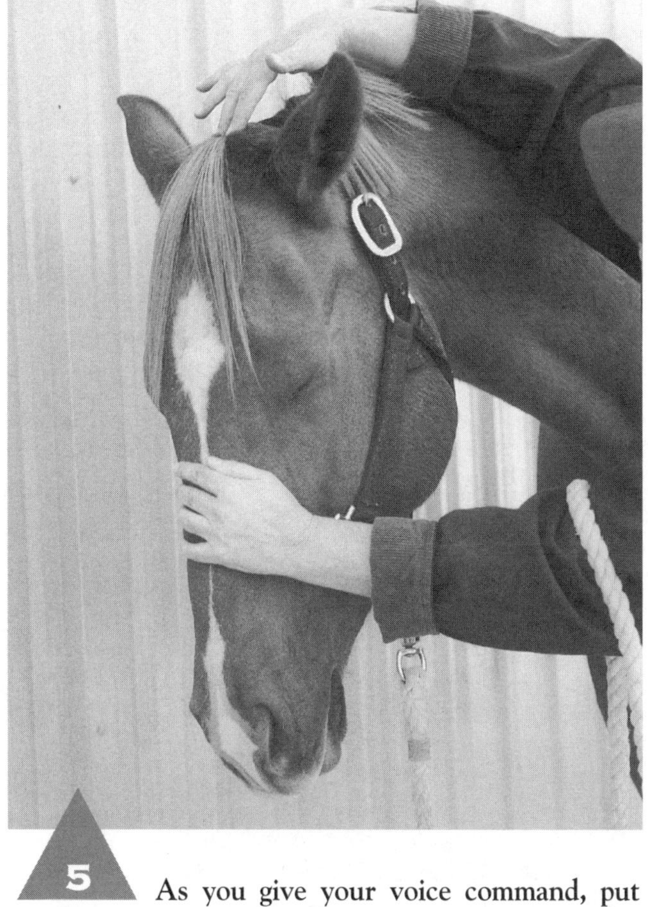

5 As you give your voice command, put fingertip pressure on the horse's poll and add light pressure with your left hand on his nose if necessary.

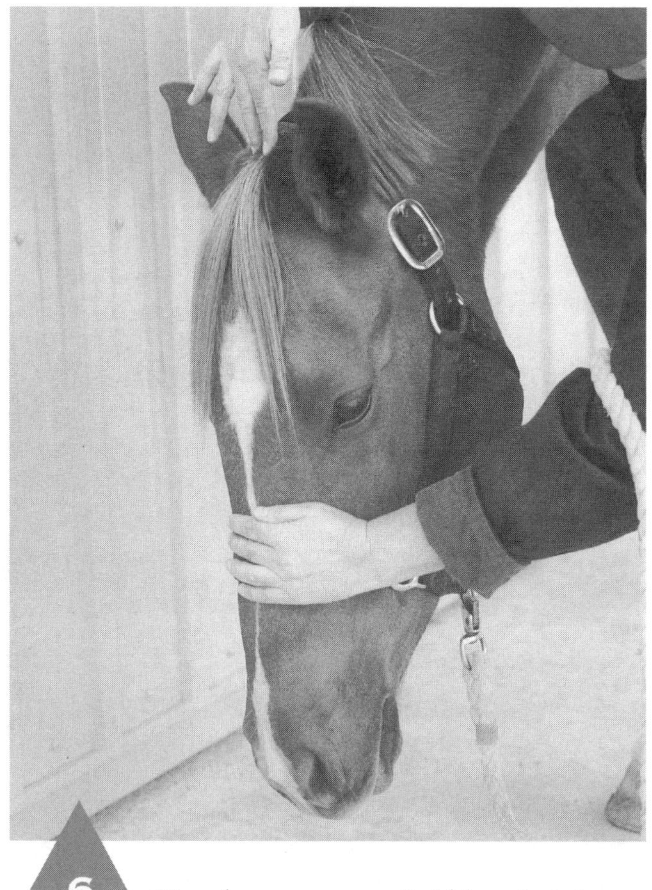

6 How low can you go? Although it is not necessary to lower a horse's head this far for most tasks, it does make a point with a horse and it tends to relax him. It also provides a good stretch for the horse's muscles.

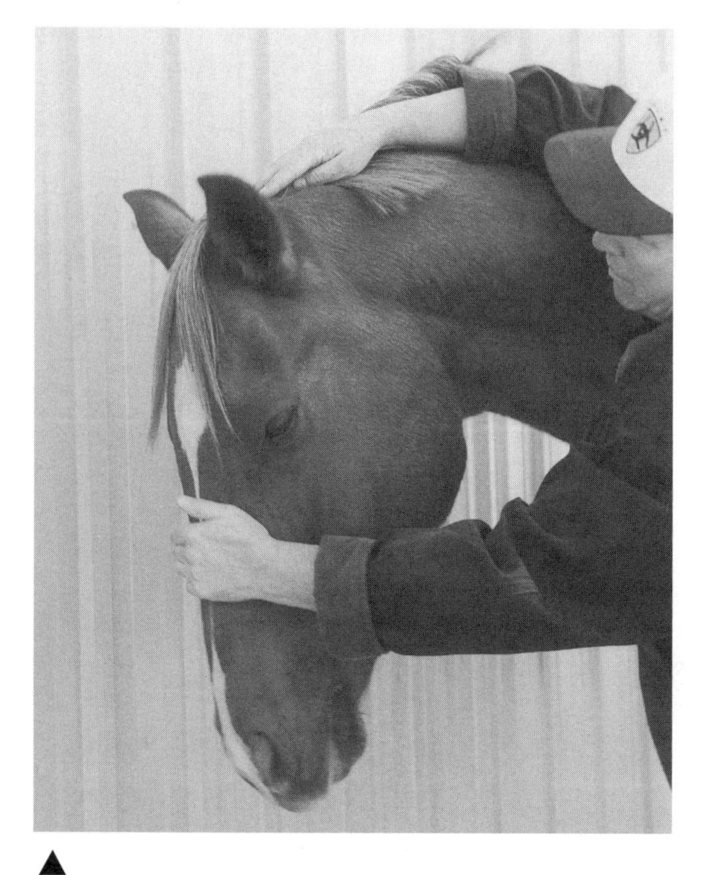

▲
PRACTICE WITHOUT A HALTER
When your horse has learned the lesson with the halter, repeat the procedure without a halter.

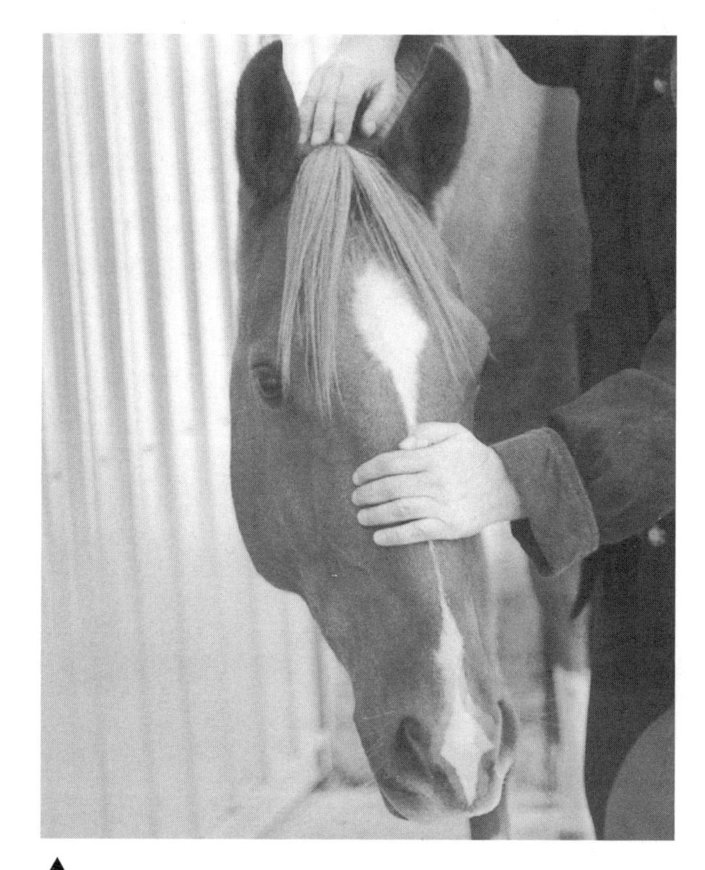

▲
ENJOYING THE LESSON
This horse, over 16 hands tall, could easily keep his head out of a person's reach. Not only has he learned his lesson but he also enjoys it.

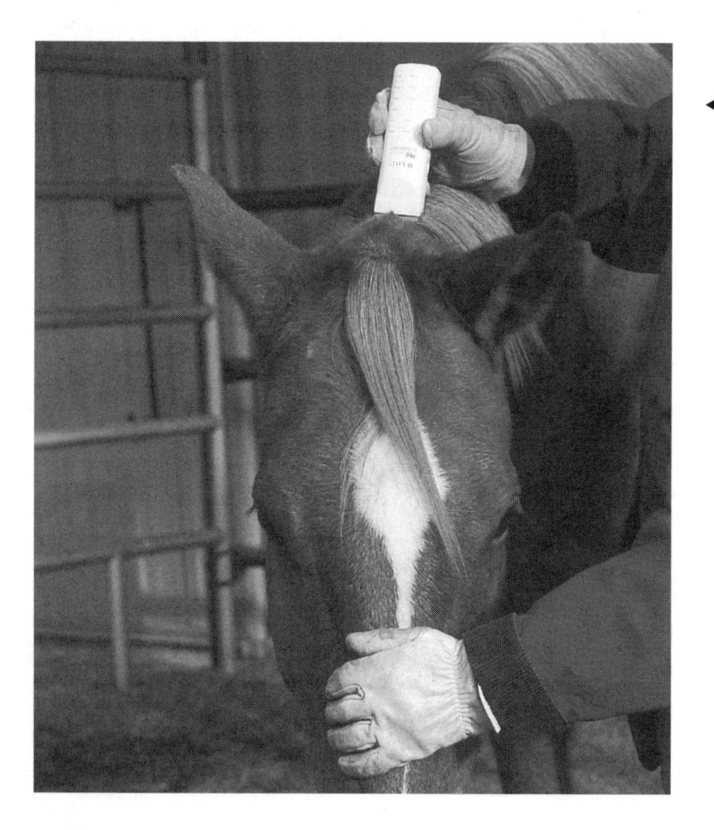

◀ **THE BENEFIT OF THE LESSON**
A benefit of this lesson is evident when you need to clip your horse's bridle path. The "head down" lesson has paid off here and will in many other situations to come.

USING A TWITCH TO SUBDUE A HORSE

A twitch is used to subdue a horse, especially during veterinary work such as treating wounds. Despite appearances, it is not the chain of the twitch that controls the horse so much as the horse's own chemicals. When a twitch is applied, it causes endorphins *(natural painkillers) to be released into the horse's bloodstream, which calms the horse. A shoulder twitch works in a similar fashion. Twitching a horse by the ear is not recommended because it tends to make a horse "head shy." Ear twitching is based on control by pain and may make a horse more excited rather than more relaxed.*

Practice applying a chain twitch when your horse is calm so you'll be very familiar with the procedure during an emergency, when your horse might be excited.

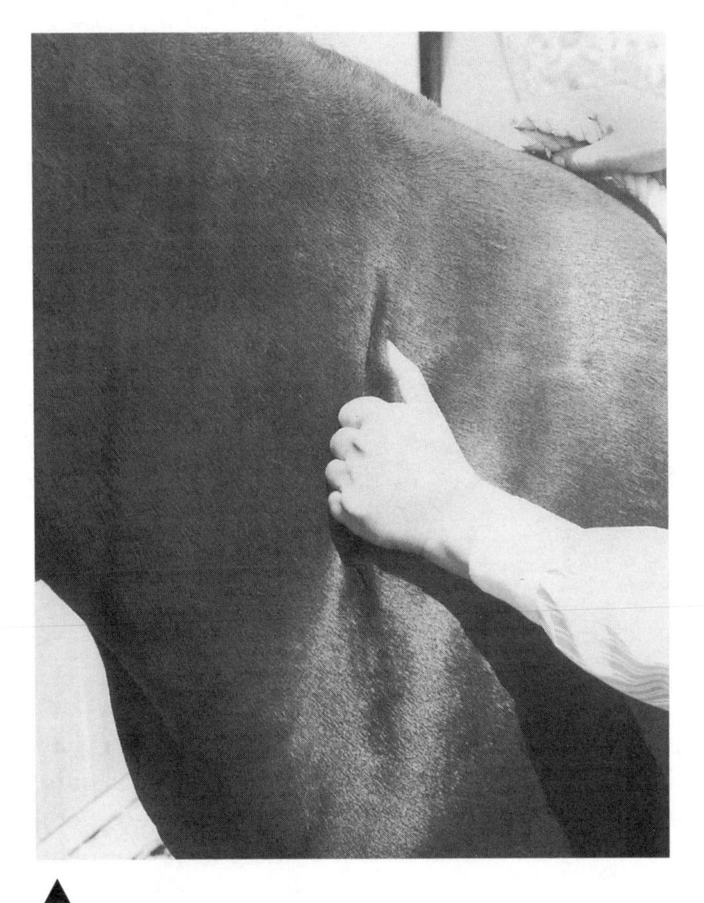

▲
SHOULDER TWITCH
A quick form of twitching that doesn't require any equipment is the shoulder twitch. Grasp a fold of skin in the shoulder area and roll your hand forward.

▲
THE CHAIN TWITCH
A conventional chain twitch is a chain loop fastened to a wooden handle.

Applying a Twitch

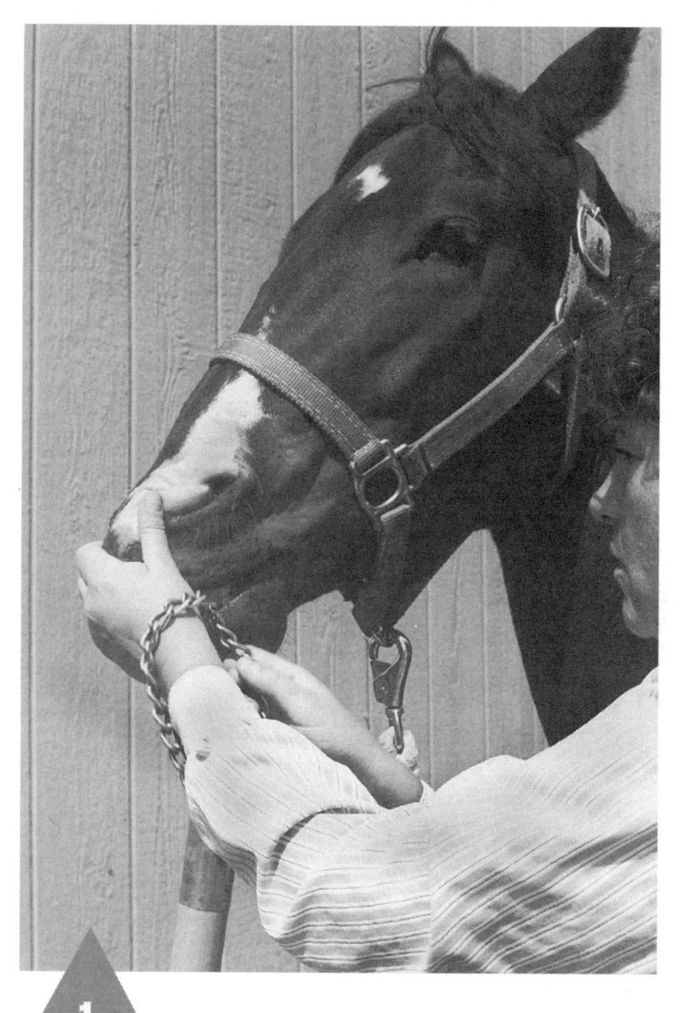

1 To apply a nose twitch, stand on the near side with your horse haltered. Put the chain loop of the twitch around your left wrist like a bracelet. Grasp a generous handful of nose tissue with your left hand.

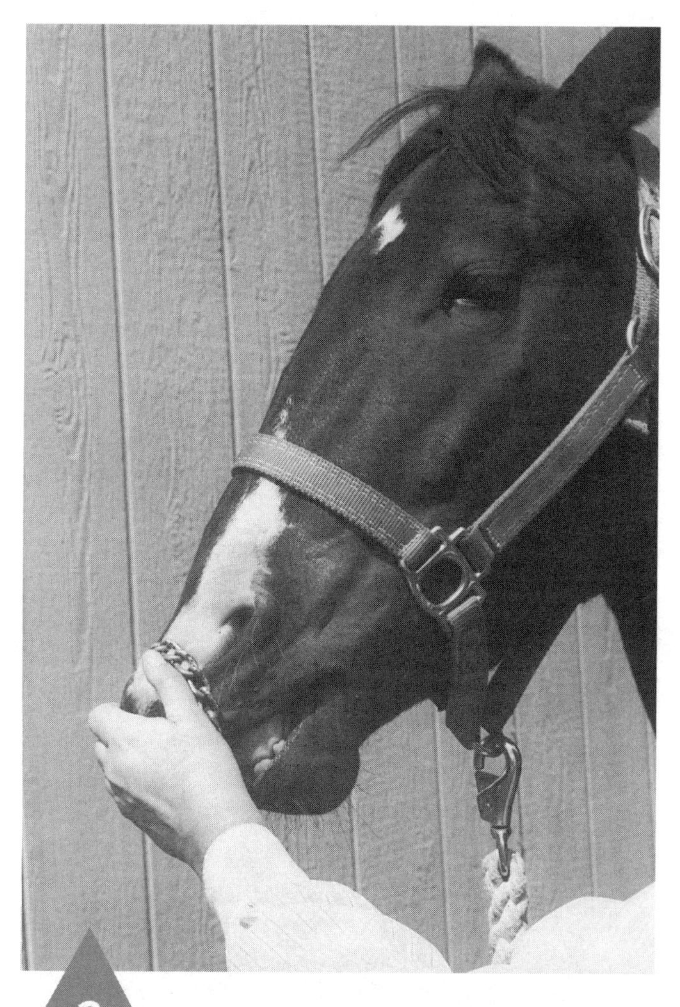

2 While you hold the nose tissue with your left hand, slip the chain loop off your wrist and onto the horse's nose with your right hand. Begin tightening the chain around the horse's nose with your right hand by twisting the handle of the twitch. (If the chain is attached in a customary fashion, twist clockwise. If the chain locks and buckles, you'll have to twist counterclockwise.)

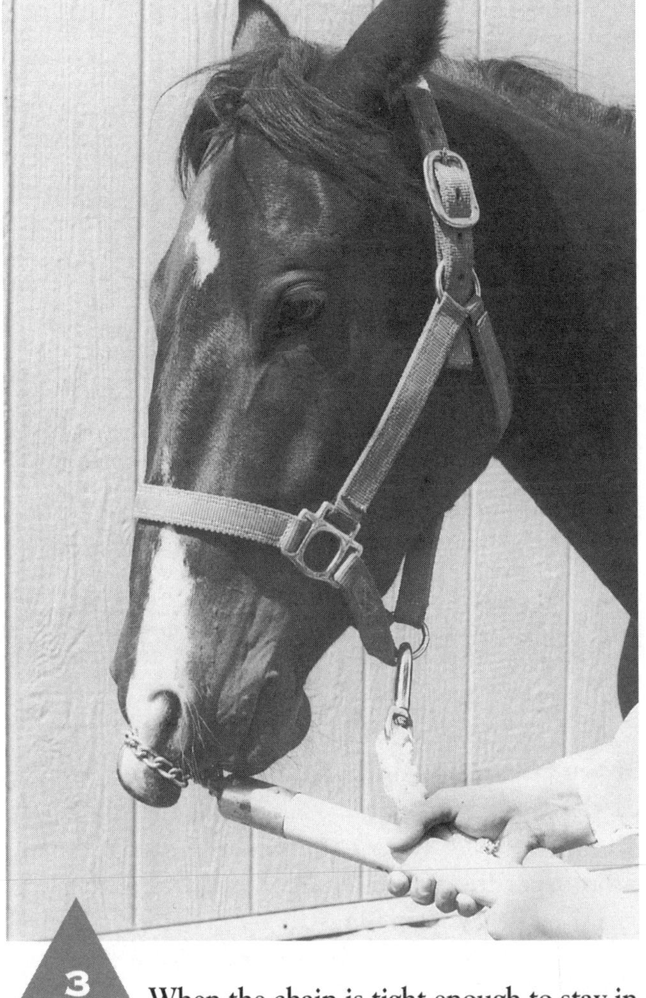

3 When the chain is tight enough to stay in place on the horse's nose, hold the twitch and lead rope together with both hands.

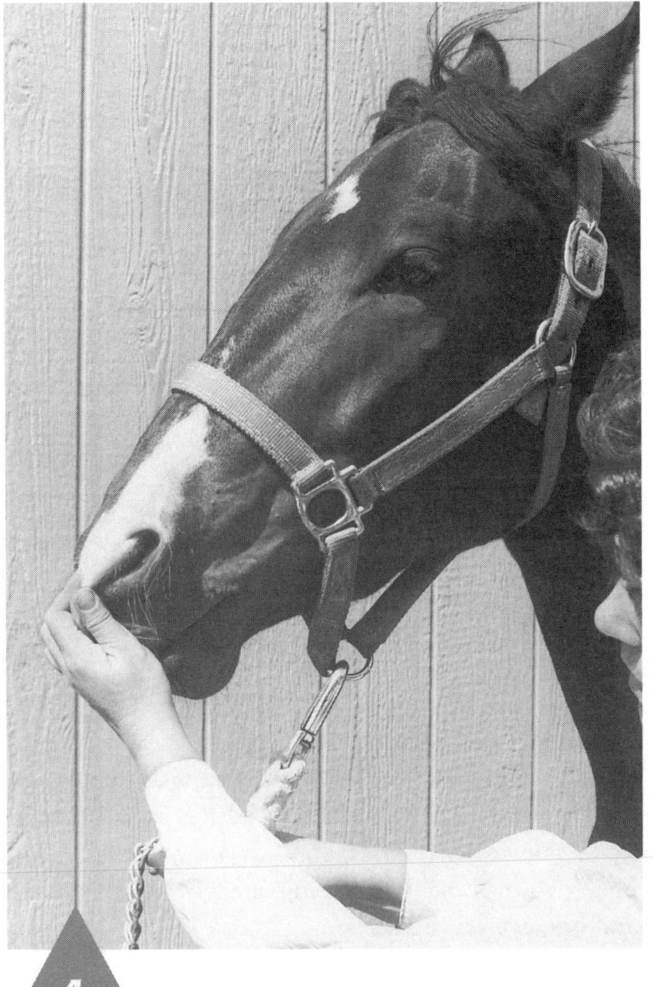

4 When you remove the chain, massage the horse's nose to restore circulation and to prevent your horse from becoming nose shy.

Holding a Horse for the Farrier

If you or your farrier prefer not to tie your horse when he is being shod or if there is no safe place to tie him such as at a horse show, you should learn how to hold a horse safely and effectively for farrier work.

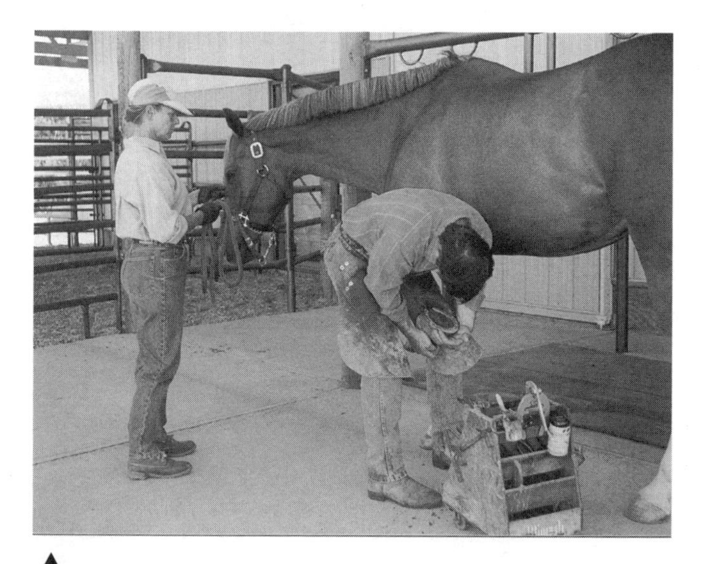

▲

Holding a Horse along a Wall
If you are holding the horse along a wall or rail and the farrier is working on a front leg, stand on the same side as the farrier, next to your horse's head, facing your horse but keeping one eye on your horse and one eye on the farrier.

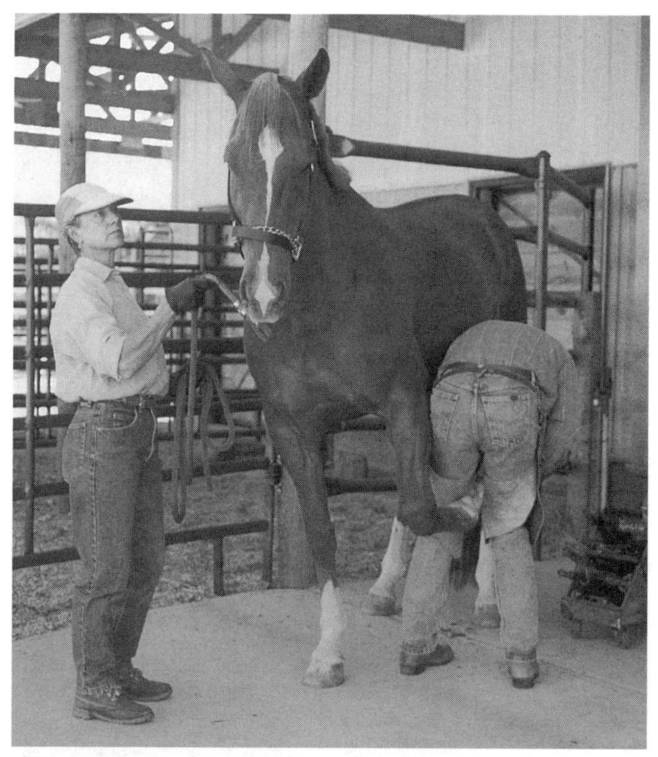

▲

Holding a Horse out in the Open
If you are holding the horse out in the open and the farrier is working on a front leg, stand on the opposite side of the horse. If the horse starts moving away from the farrier, you can touch the horse on the shoulder to keep him still. Keep the horse's head up and still to make your farrier's work easier.

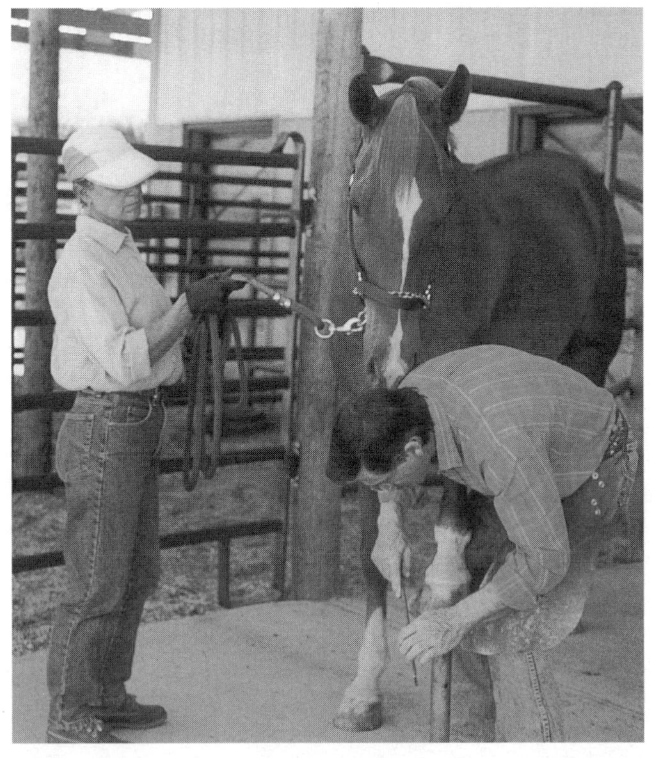

▲ **KEEP AN EYE ON THE HORSE**
When the farrier brings the horse's leg forward on the hoof stand, don't watch what your farrier is doing; keep an eye on your horse. Here the horse is nuzzling the farrier's back, which covers his shirt in horse slobber and is often the prelude to a bite, not a kiss. Farriers have been seriously hurt by this lack of attention on the part of the handler.

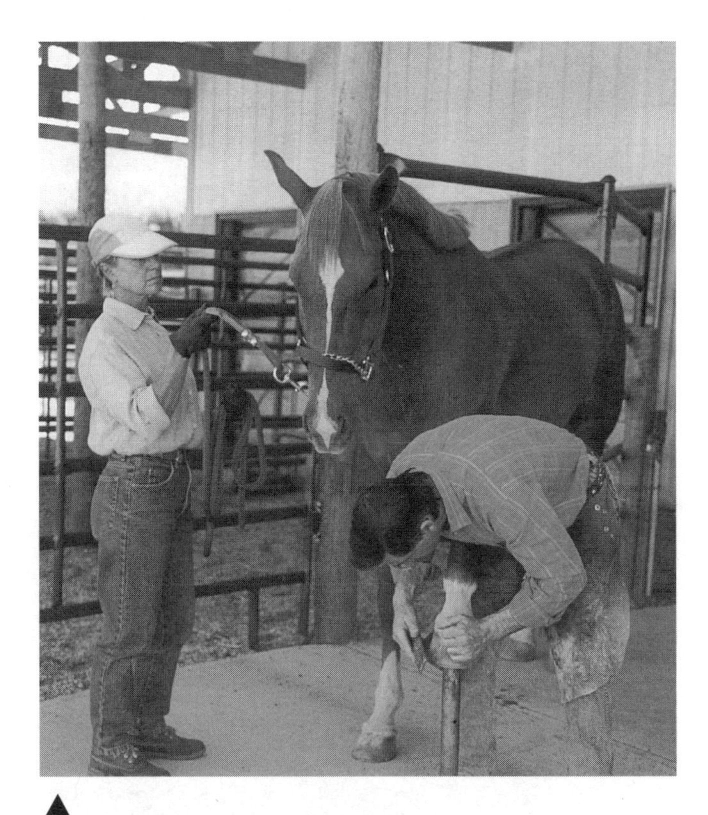

▲ **KEEP THE HORSE'S HEAD UP**
Make sure you keep the horse's head up and away from the farrier's back.

◀ **KEEP THE HORSE'S HEAD FORWARD**
When the farrier is working on a hind leg and the horse is along a rail or wall, keep the horse from turning around to look at what the farrier is doing. This only makes the farrier's work harder, as he supports the shifting weight of the horse.

Help the Horse to Balance

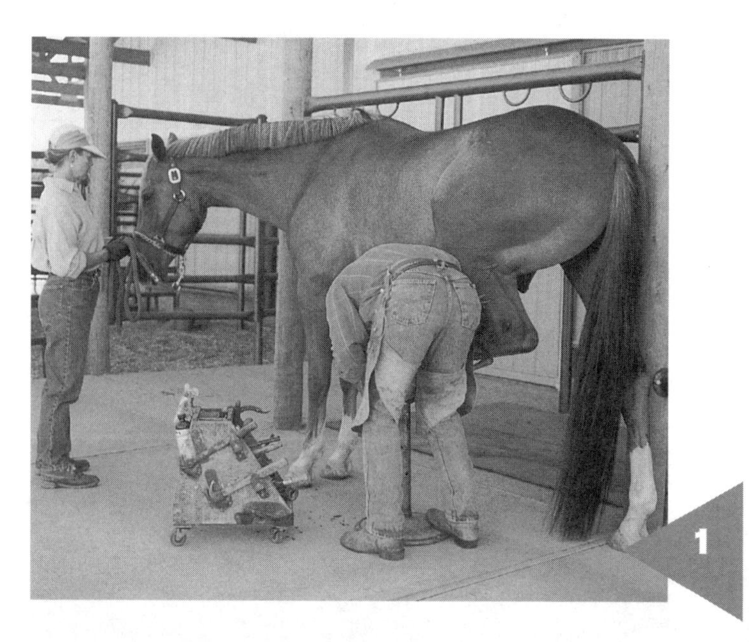

1 When the farrier brings the horse's hind leg forward on the hoof stand, let the horse lower his head somewhat; this helps him to balance in this position.

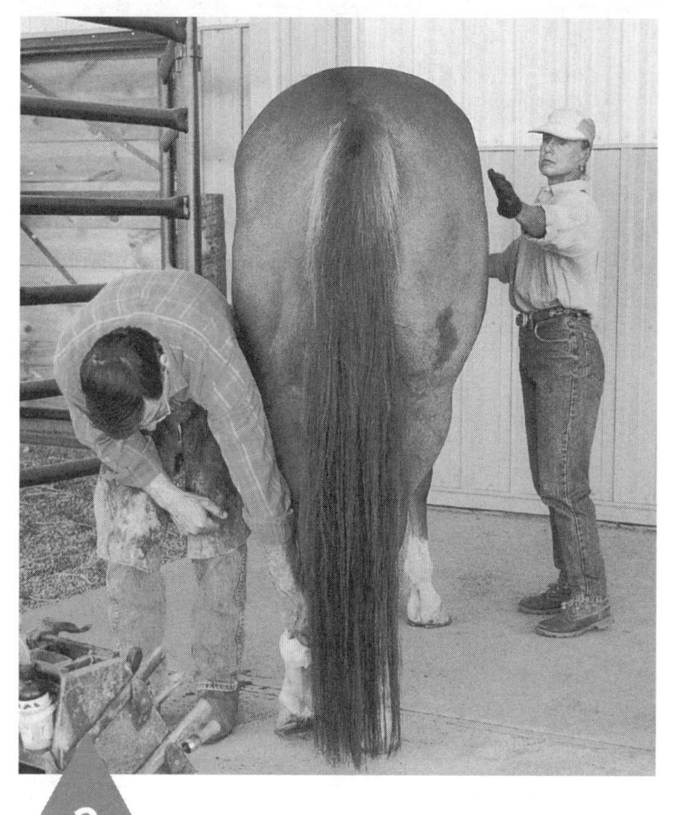

2 When the farrier is getting ready to pick up a hind foot and you are holding the horse out in the open, the horse may try to move sideways.

3 Prevent the horse from moving sideways by having your hand ready to correct the horse if he begins to move.

Picking Up a Horse's Feet

Your daily hoof checks and most farrier work are usually performed on a horse that is tied. The horse must be standing balanced, not right up against a wall or post.

Picking up the Front Foot

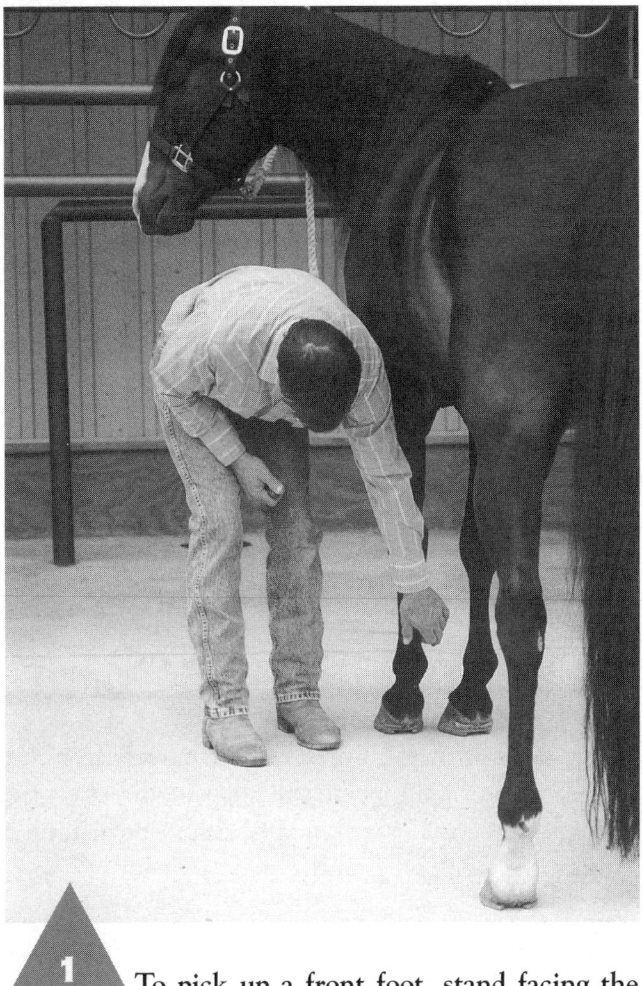

1 To pick up a front foot, stand facing the opposite direction from the horse. Use a voice command such as "Foot."

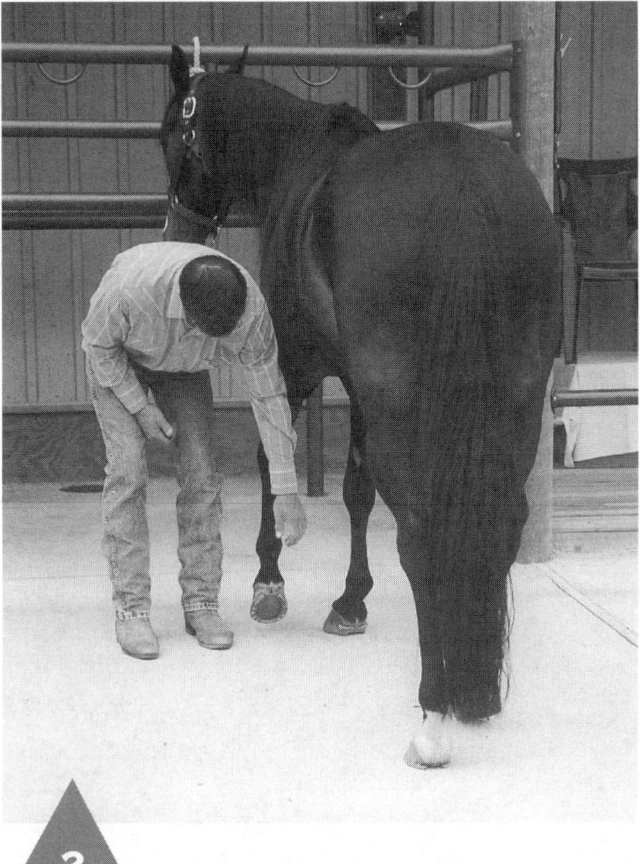

2 As you pick up a front foot, use your index finger and thumb to pinch the tendon area above the fetlock to signal the horse to pick up his foot. Never try to lift up a horse's foot. It is simply too hard and may teach a horse to resist you.

How strongly you must pinch the tendon area will depend on the horse and his level of training. When first training a horse to pick up his feet, you might have to pinch harder. As the horse learns what you want, just a light touch will suffice. As the horse begins lifting his foot, let go of the tendon area and be ready to catch the foot when it is in the air.

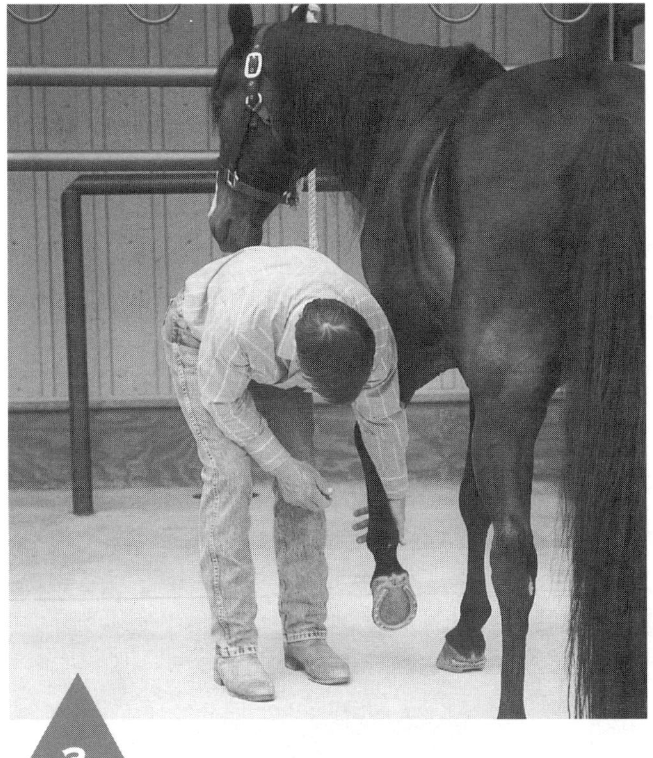

3 Grasp the front of the cannon with the palm of your hand.

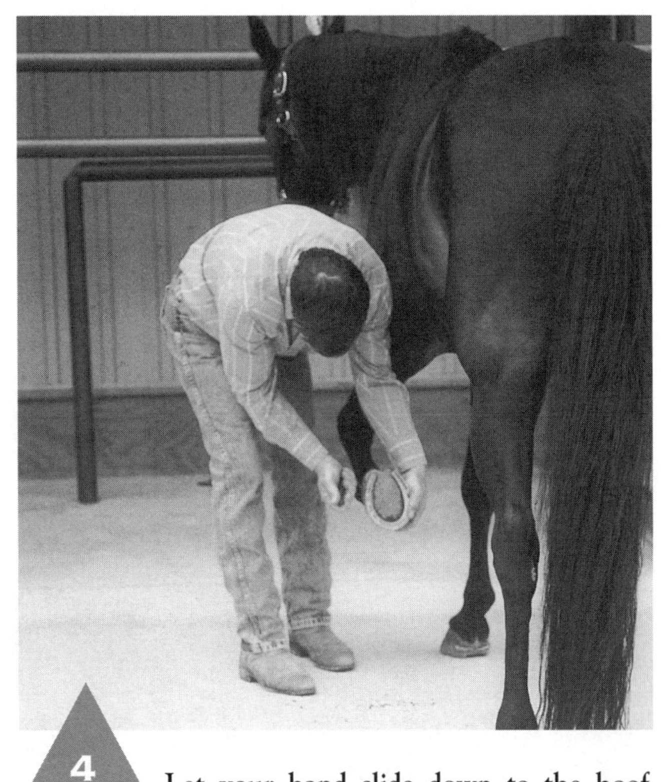

4 Let your hand slide down to the hoof where you will support the leg as you clean the hoof.

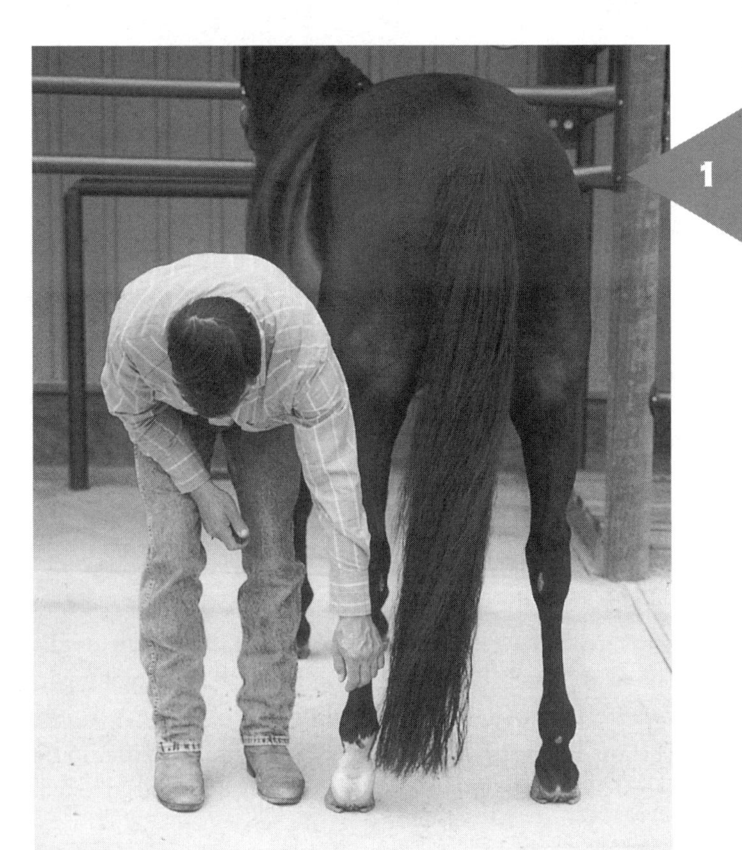

Picking Up the Rear Foot

1 The same principles apply when you want to pick up a rear foot. Stand facing the opposite direction from the horse. The horse should not move or pick up his foot until you give him the cue.

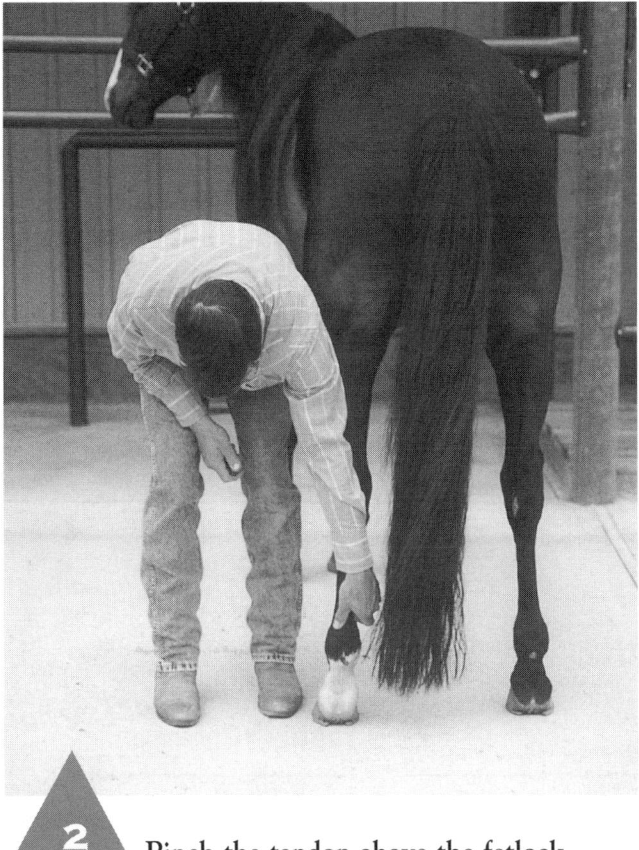

2 Pinch the tendon above the fetlock.

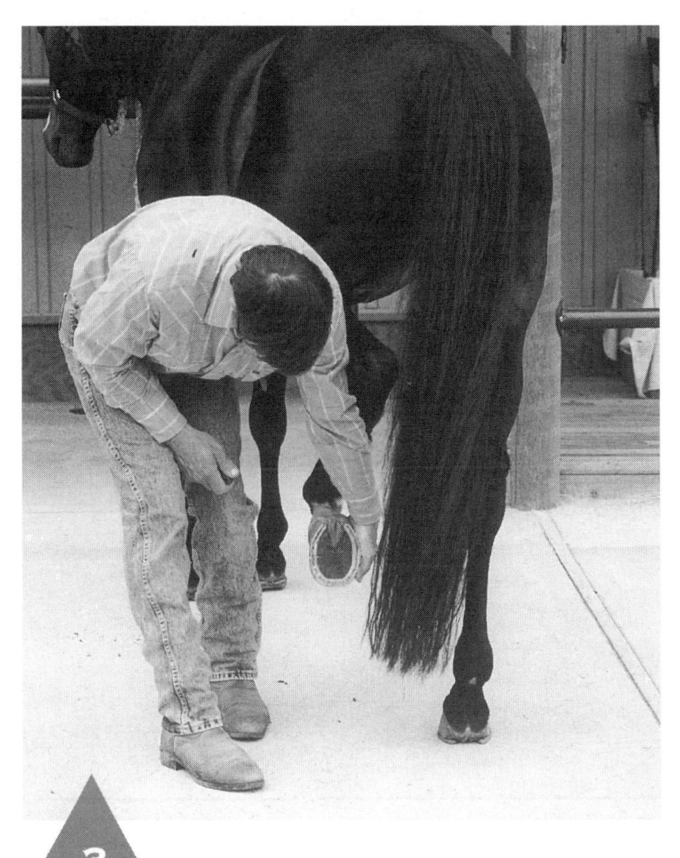

3 Cradle the hoof in the hand closest to the horse.

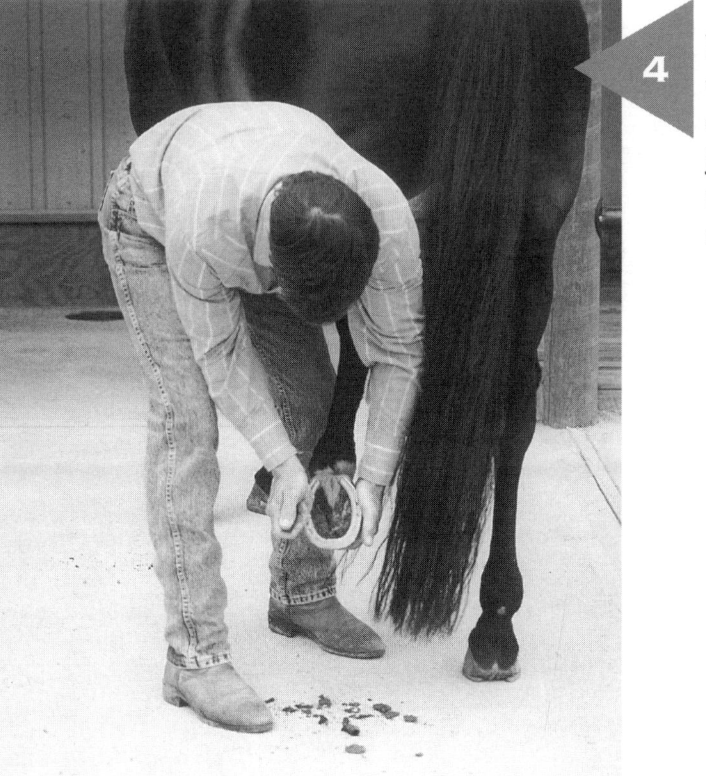

4 Bring the hoof back slightly and rest it on or against your leg. Keep the horse's leg low while you work on it. This makes it most comfortable for the horse's joints. If you practice hoof-handling skills in a systematic manner, you will end up with a cooperative, relaxed horse in any situation.

Grooming Techniques

Grooming your horse serves a number of purposes. It removes dirt, sweat, glandular secretions, dead skin cells, and hair. It facilitates shedding and brings natural oils to the surface. The thoroughness of the grooming allows you to make a close inspection of your horse's skin, head, mane, tail, legs, and hooves. The physical benefits are quite evident; grooming can help you monitor the health of your horse, produces cleaner skin and hair, and adds a glossy sheen to your horse's hair coat. A valuable result of this is cleaner tack, and thus less tack maintenance.

But the advantages extend even further, for both the horse and you as the owner. Grooming provides mental and physical preparation for the work to come, and so is a valuable warm-up. The massage increases circulation and relaxes the horse's psyche. It accustoms a horse to being handled and helps to desensitize ticklish areas. Grooming presents an opportunity to work on various horse skills such as patience and obedience, and allows you to spend hands-on time with your horse.

Storage Area

This grooming equipment storage area is located near cross ties and has cubicles for grooming totes that are filled with brushes, combs, curries, and sponges. It also features shelves and brackets for sprays and other grooming supplies. There are also several electrical outlets for plugging in clippers and a vacuum. An area such as this keeps equipment handy, organized, safe, and clean.

GROOMING EQUIPMENT

◀ COMMON GROOMING TOOLS
These are some of the most common grooming tools. Lower row from left to right: round rubber curry, rectangular rubber curry (my favorite); grooming gloves, cloths, hoof pick. Next row: combination sweat scraper and shedding blade. Next row: dandy brush, body brushes (three types), grooming mitt finisher. Top row: apron with tools, apron with spray bottles.

◀ ORGANIZE YOUR TOOLS
A plastic tote helps you organize all your grooming tools so you can move them with you from one location to another.

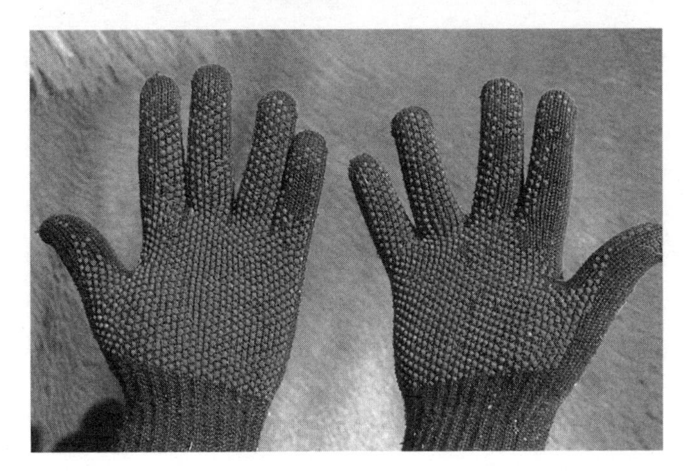

◀ GROOMING GLOVES
Knit grooming gloves with rubber dots not only keep your hands clean but also are perfect for hand rubbing the head and legs.

Grooming Procedure

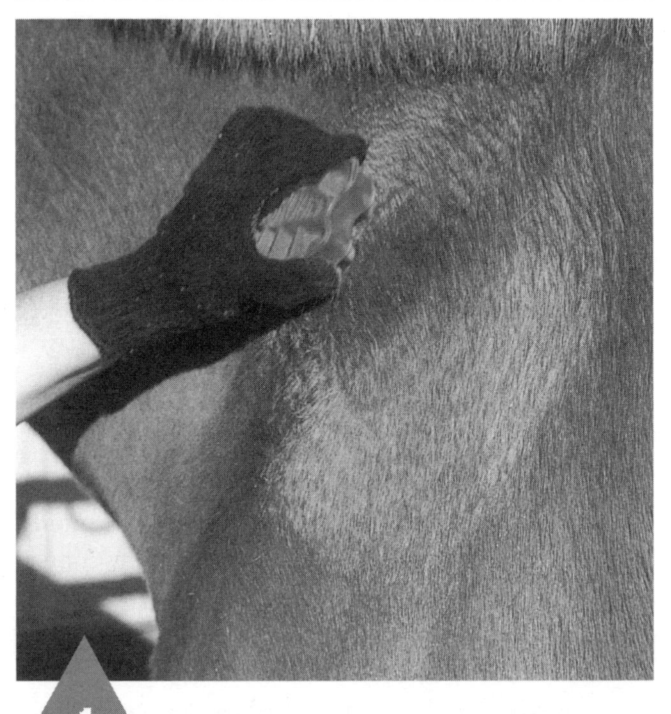

1 Begin body grooming with a vigorous circular motion using a rubber curry. This rectangular rubber curry fits smaller hands particularly well and is not as fatiguing to hold as some of the round curries.

2 Then use a stiff-bristled dandy brush to remove the loose hair, dirt, and scurf from the horse's coat. Start each stroke with the brush flat on the horse's coat.

3 With a quick flick of the wrist, whisk the dirt into the air.

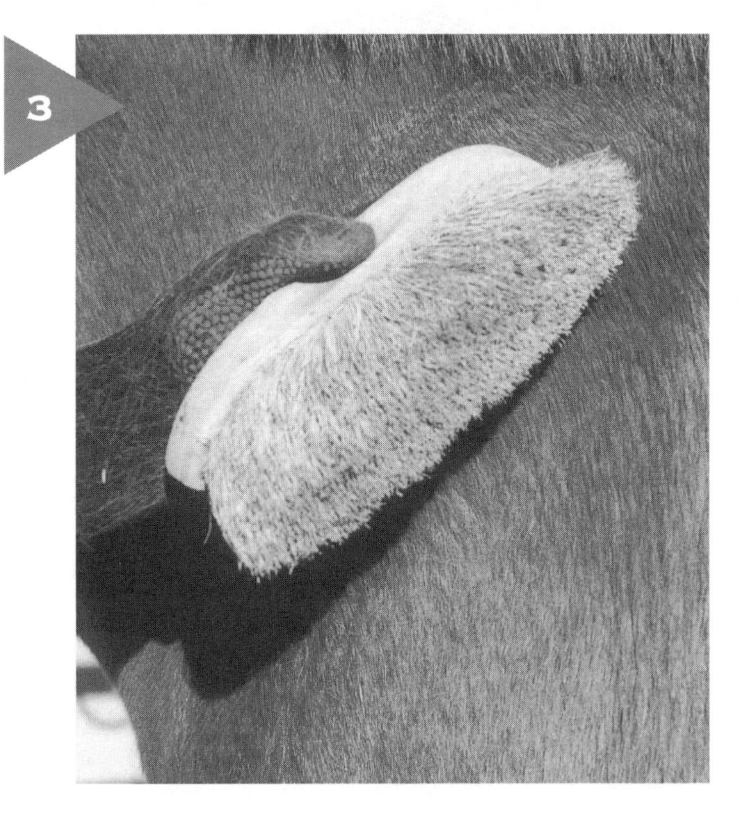

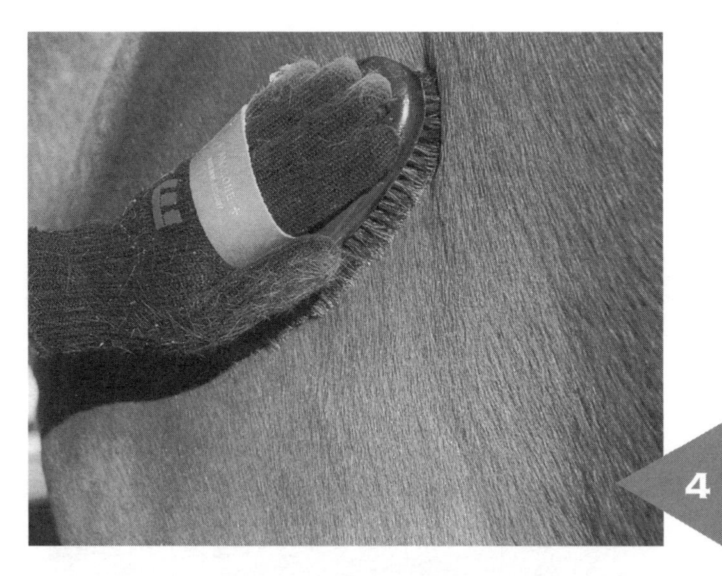

4 Next, with the body brush of your choice, brush in long strokes in the direction of the hair growth to further clean the coat.

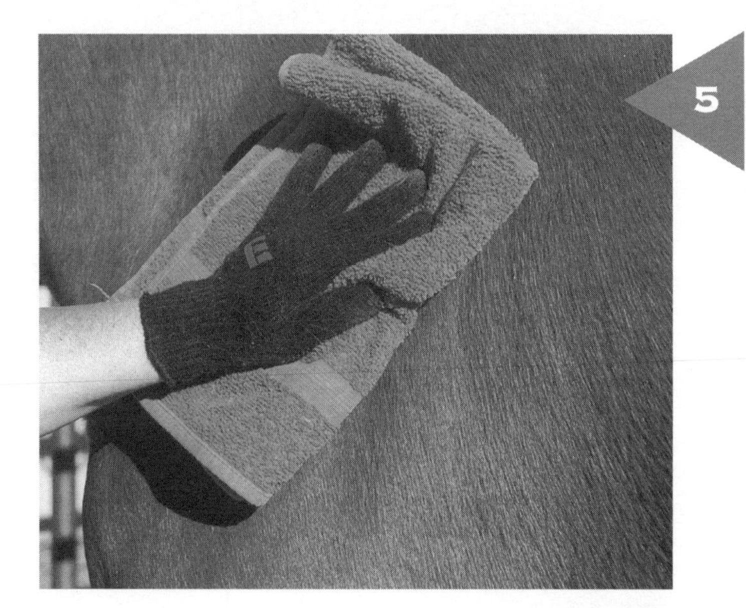

5 To give the hair its final cleaning and to set the coat, spray water or a diluted skin bracer onto a clean terry cloth. I use 1 part liniment or skin bracer to 10 parts water. (After riding your horse you can use this same solution to wipe away saddle and bridle sweat marks.) Wipe the coat in the direction of hair growth. The dampened cloth makes a great dirt magnet.

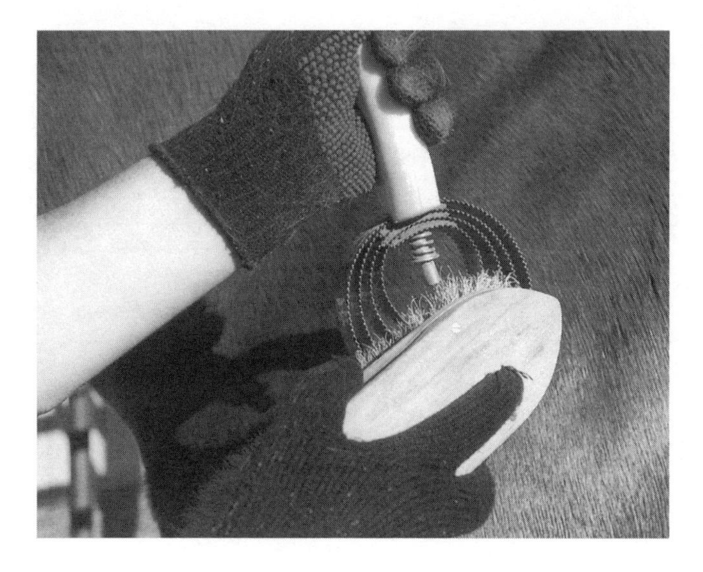

◀ CLEAN YOUR BRUSH

Periodically clean your dandy brush by stroking it across the metal curry comb. The metal curry comb is not to be used on the horse's body. The sharp teeth could scratch the horse's skin and set the stage for skin problems.

Grooming the Head

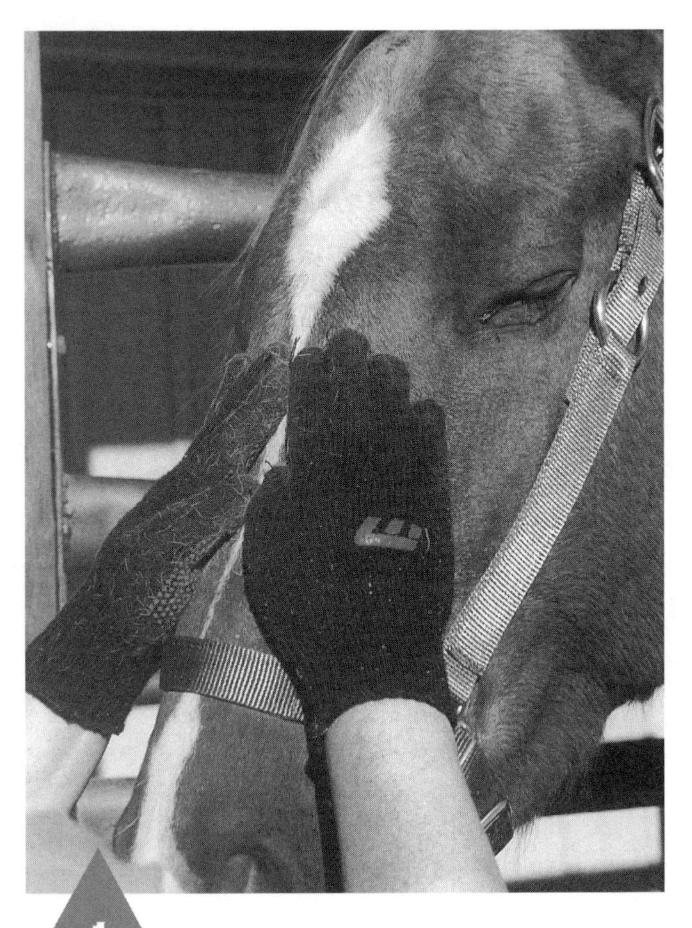

1 Use your rubber-dotted grooming gloves in place of a curry and vigorously rub the horse's head all over. You can see how relaxing this is for the horse. This horse is wearing a grooming halter.

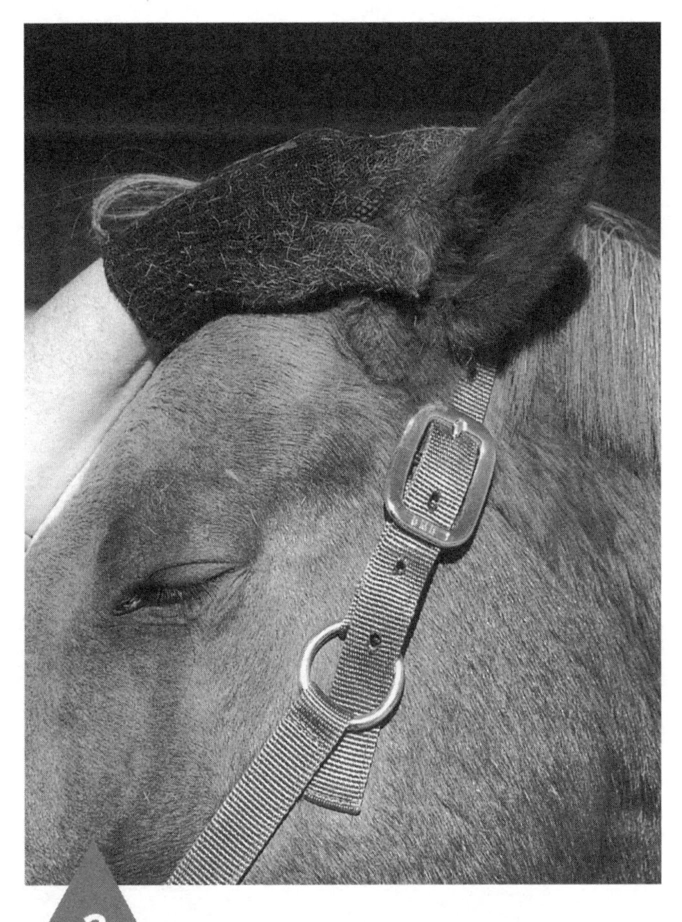

2 Don't forget the ears. Contrary to what some people think, most horses like their ears handled and melt into a puddle of contentment when given a good ear rub, inside and out.

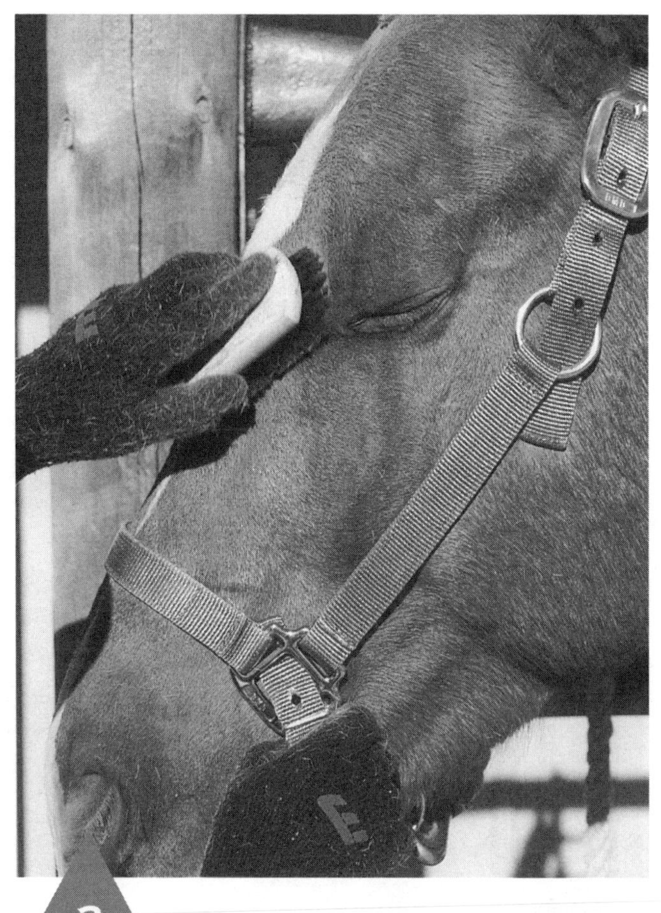

3 Use a soft-bristled brush to whisk away the hairs and dirt that your rubbing has brought to the surface, taking care around the eyes.

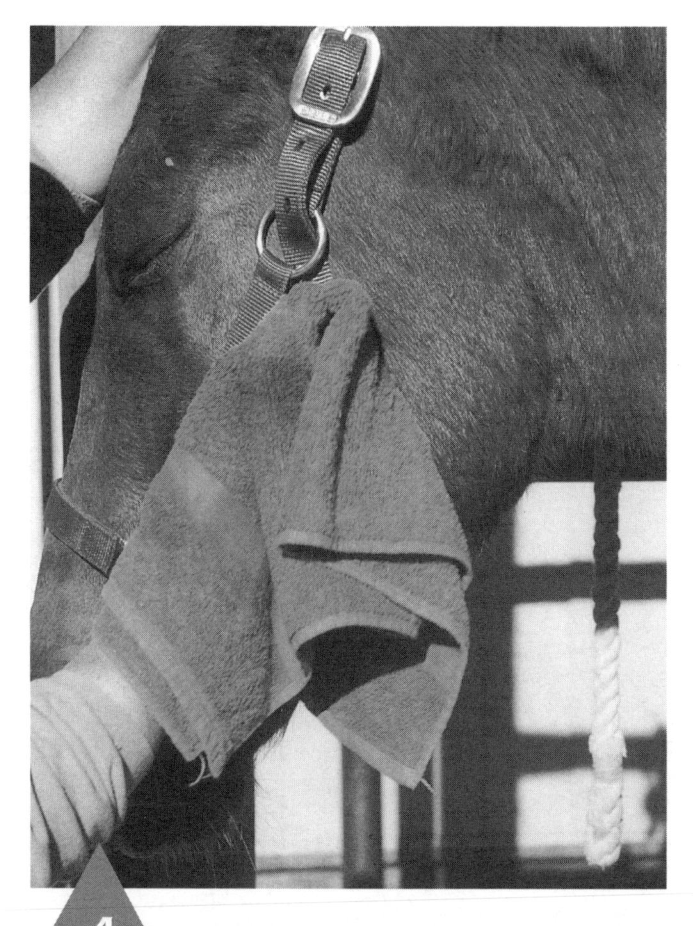

4 After brushing, use your dampened grooming cloth to wipe the entire head.

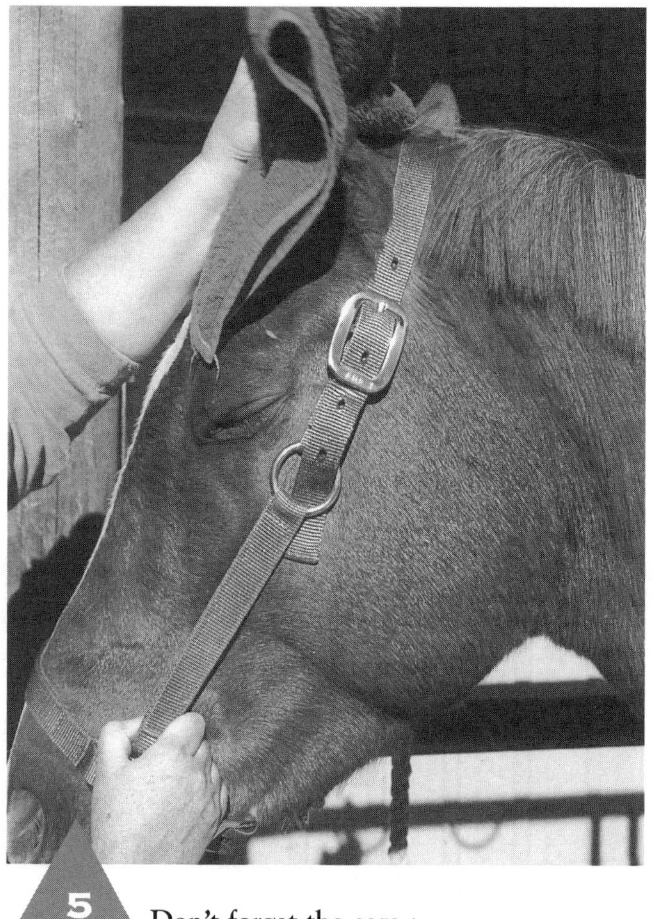

5 Don't forget the ears.

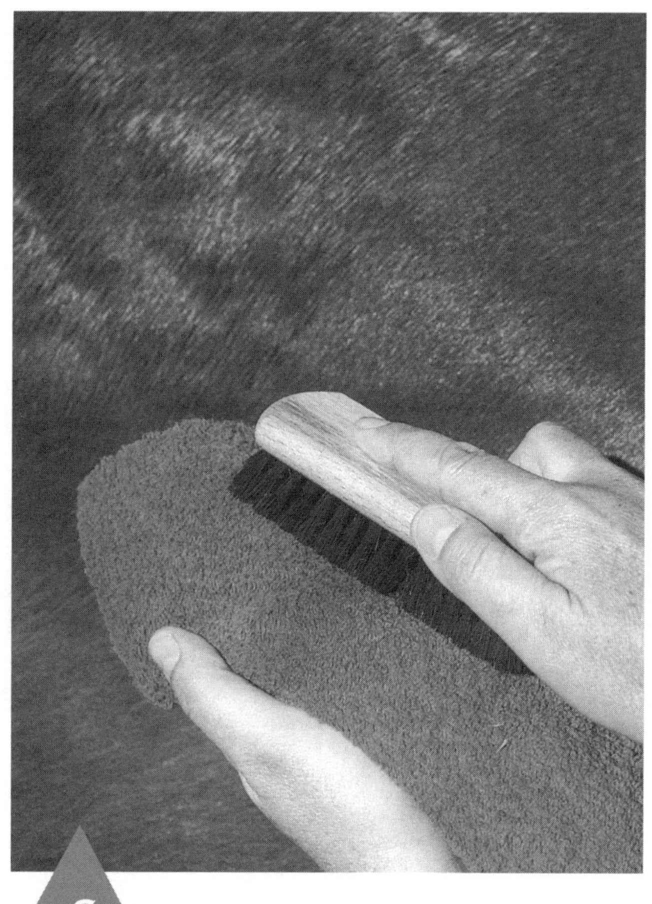

6 To finish the head and set the hairs, you will need to use the face brush once again. To clean the face brush that you previously used, run the bristles across the dampened grooming cloth a few times.

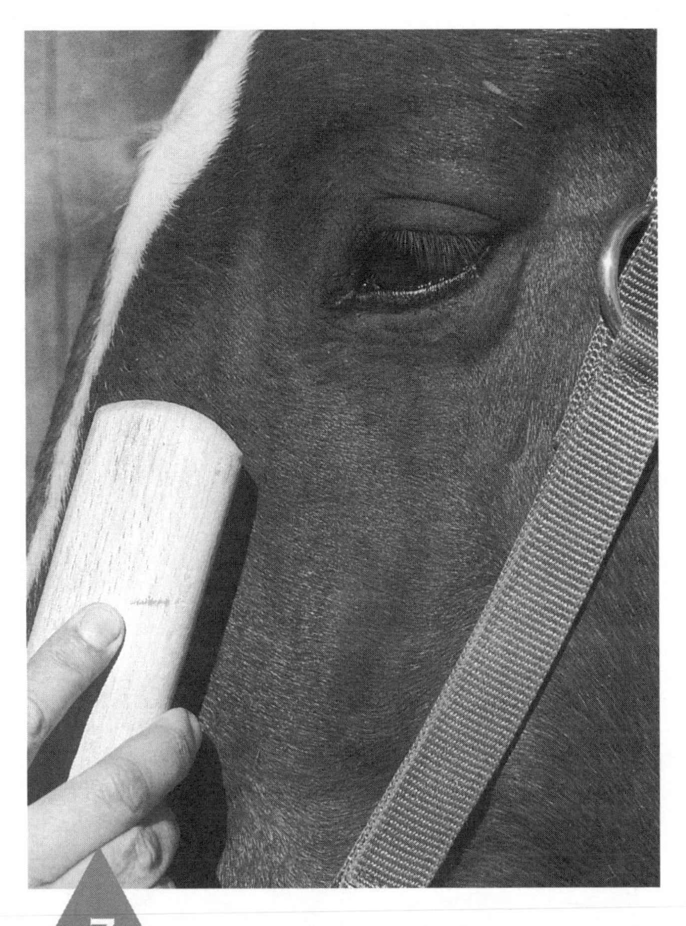

7 Finish grooming the head by carefully brushing the horse's face, cheeks, throat, and poll area in the direction of hair growth.

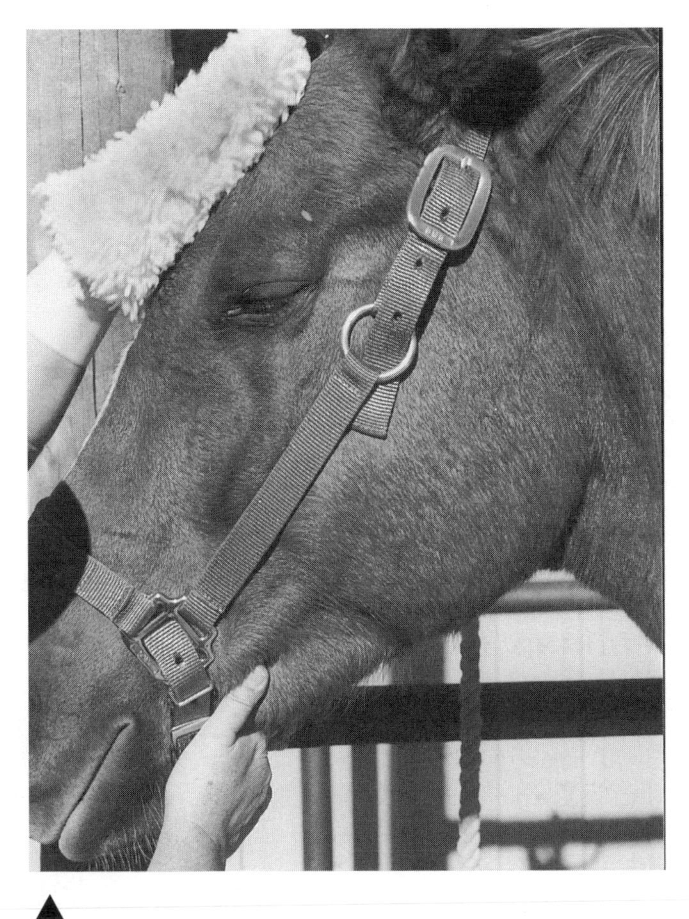

▲ GROOMING MITT
An alternative to the terry cloth is a cotton grooming mitt.

MAKING AND USING A WISP

A wisp can be simply a handful of straw used to groom a horse or a woven pad of grass hay or straw with a handle that is designed to be used as a vigorous grooming tool for the horse's body. A well-made wisp might last you a month or so, depending on how dirty or hairy the horses are that it is used on. You can use a new damp wisp to set a horse's coat.

MAKING A "HAY ROPE"

1

Break apart a flake or two of long-stemmed, soft, fine grass hay or straw into a fluffy pile on the ground. Sprinkle it generously with water.

2

Tie a 2-foot piece of twine to an immovable object such as a wall so that you end up with one 6-inch tail and one 18-inch tail. Leave the shorter tail free and start twisting a wad of hay onto the long tail beginning as close to the wall as possible. Oppose the action of one hand with the other so that the twist stays tight. Continue adding small amounts of hay every inch or so, twisting and gently pulling to fashion a hay rope. Continue past the end of the twine until you have 9 feet of hay rope that is about ¾ inch in diameter. Lay another piece of twine into the working end and continue adding hay and twisting for another foot or so. Have an assistant detach the twine from the wall and help you maintain tension.

Weaving the Wisp

With the hay rope you just made, weave a two-lobed framework according to the following diagram.

1 Form a pretzel-like shape with the twine.

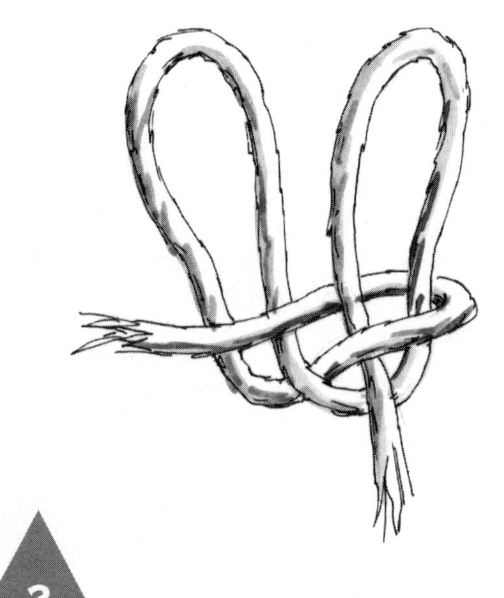

2 Weave the loose end on the right around and behind the right lobe of the pretzel.

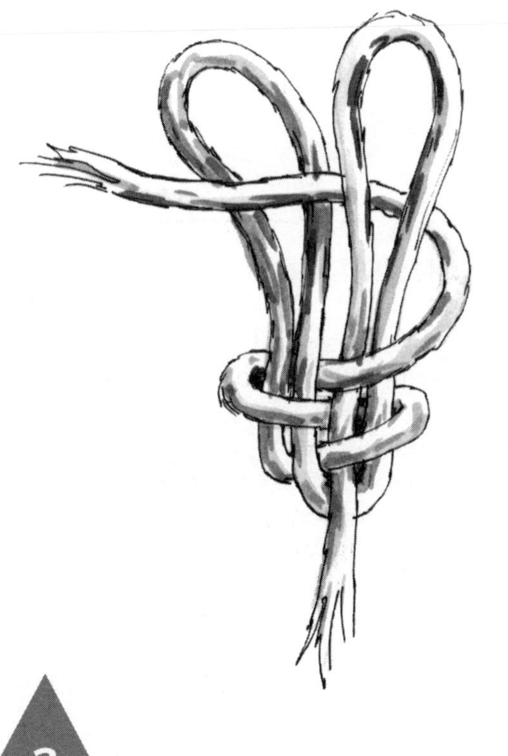

3 Wrap that same end around and in front of the left lobe, then behind it and in front of the right lobe of the pretzel.

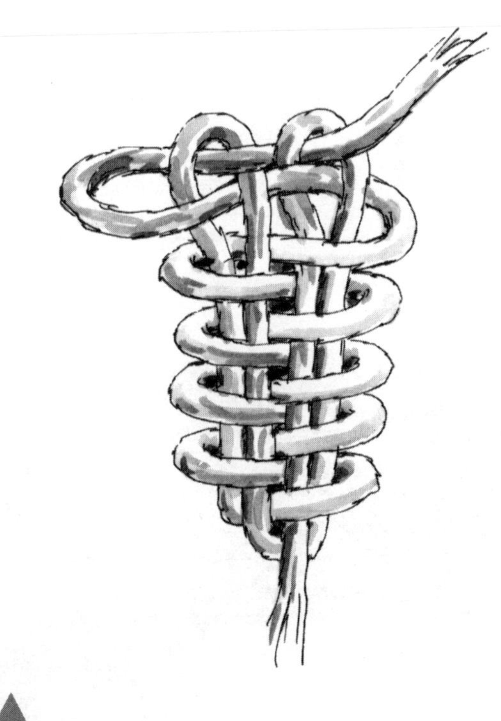

4 Continue wrapping in this fashion until you reach the top of the "lobes." Then run the loose end *through* one lobe, then the other, twice.

▲
You should end up with twine emerging from two places.

▲
Tie the two pieces of twine together to fashion a handle that fits your hand comfortably.

Using the Wisp

Use the wisp as a curry or for body stropping. For stropping, hold the wisp 6 to 12 inches away from the horse's deep muscles (such as shoulder, hindquarters, neck). Bring the wisp to the muscle smartly with a flat bang and move it briskly off the body in the direction of hair growth. This causes the horse to contract his muscles, then relax. Apply the wisp again with a flat bang. The alternating contraction and relaxation acts as a purification pump that carries nourishment in and waste products out of the muscles.

USING A VACUUM

Vacuums can be a healthy, labor-saving grooming tool. A good stable vacuum must be safe and powerful, easy to maneuver, and must operate with minimal noise and reasonable energy efficiency. It's a plus if the vacuum comes with quality grooming tools.

A vacuum is a great asset for regular grooming and a big help during spring and fall, when baths often are not feasible. During shedding times, use a rubber curry and shedding blade to remove the large amounts of hair, then follow up with a vacuum. A vacuum removes the scurf and dirt in longer, thicker coats that is difficult to dislodge by conventional grooming methods. Vacuuming a shedding horse keeps clothing, saddle pads, and horse blankets cleaner. Vacuuming horse hair off clothing and equipment stretches the time between laundering and makes it easier on the washing machine. Vacuuming a horse removes dirt, dander, and scurf and minimizes the amount of debris that is in the air, making the grooming environment much healthier for both the groom and the horse.

NOTE: *When one end of a vacuum is sucking, the other end is blowing air out of an exhaust tube. If the barn floor is dirty, dusty, or hair covered, all of that debris will be sent back into the air by the exhaust. That's why a built-in model is much healthier for both the horse and the groom. The exhaust is vented to the outside of the building.*

CAUTION: *With portable vacuums, be aware that the electric cord and the hose are underfoot and can be stepped on or can cause the groom or horse to trip.*

Never use a vacuum to suck up water or to vacuum a wet animal. In fact, if you try to vacuum a wet horse, you could seriously shock yourself or the horse in the process. If a horse is wet, dry him thoroughly before vacuuming. Some vacuums have blower features for drying wet horses.

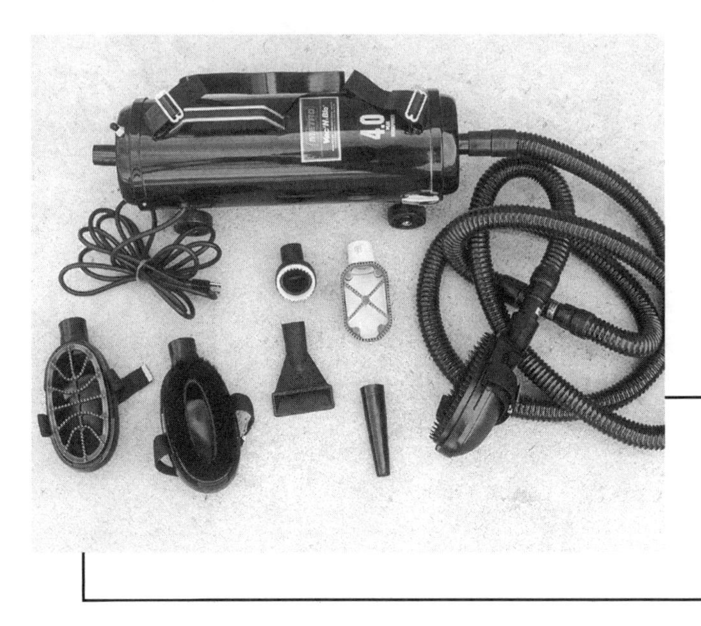

THIS IS A canister-style horse vacuum on wheels with an extra-long hose and grooming tools.

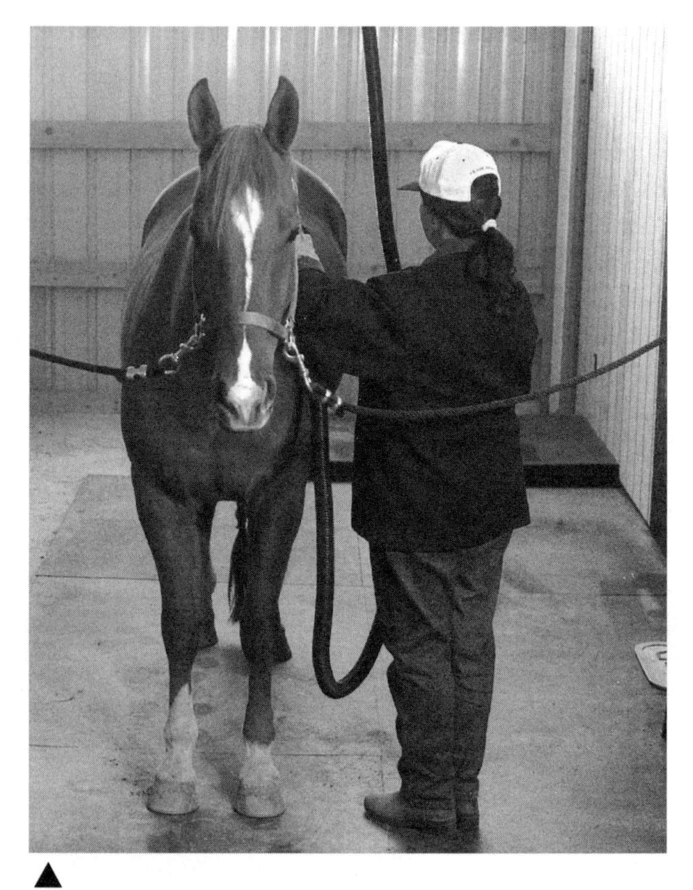

▲
Overhead Mount

Mount central vacuums on a wall away from the grooming area so only the overhead vacuum hose is in the area where you'll be working. This is much quieter and safer with no electric cords or hoses underfoot. In addition, the exhaust is vented outside the building.

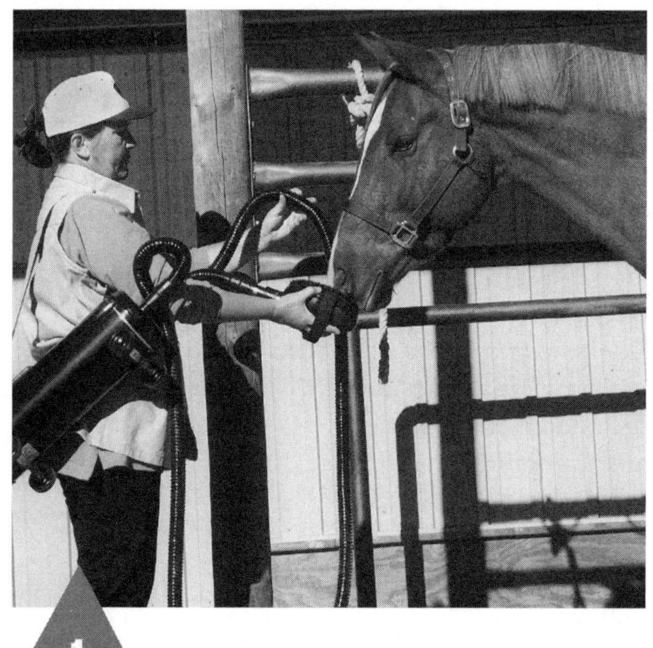

1 Before you groom your horse with a vacuum for the first time, let him inspect it by sight and smell.

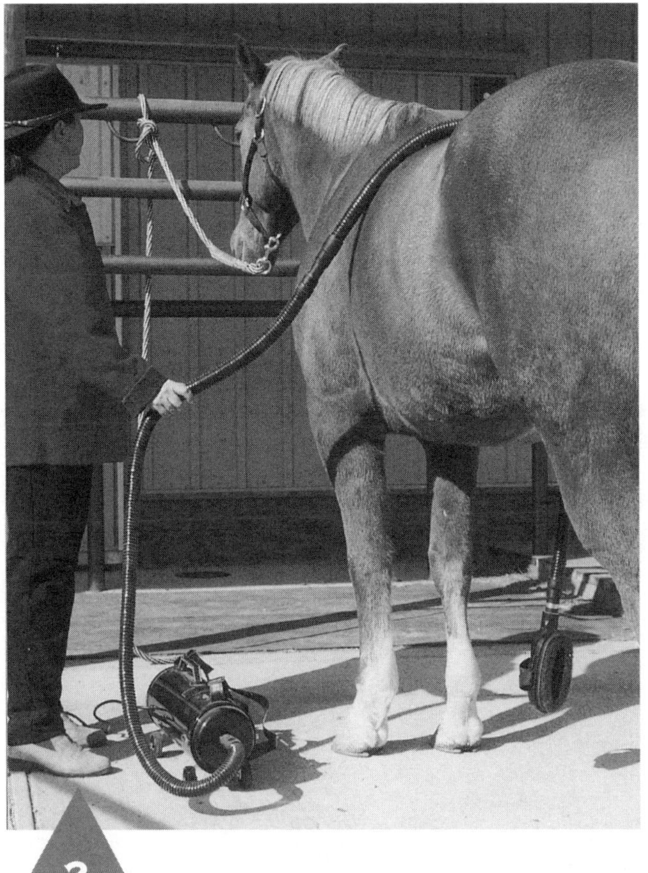

2 Accustom him to the hoses by passing them over his back.

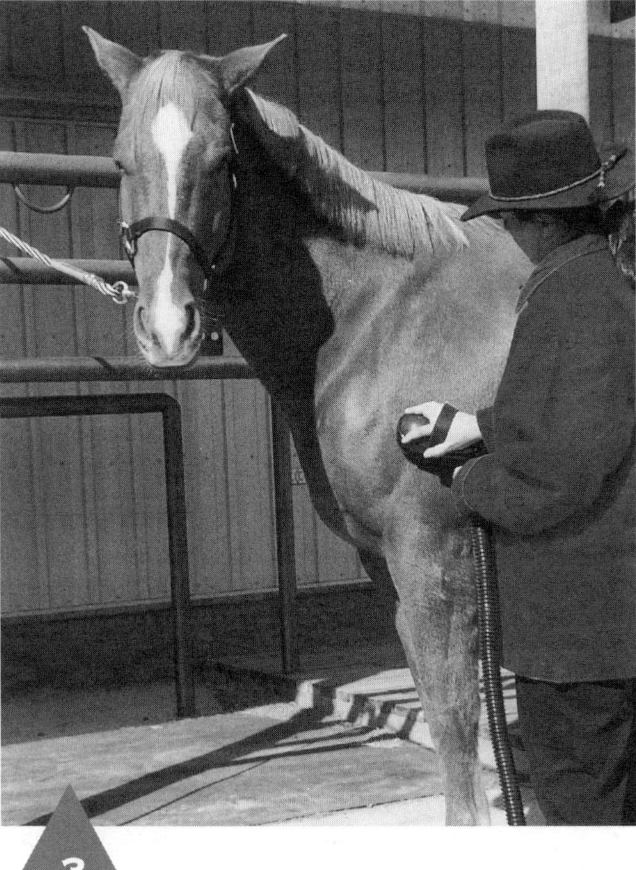

3 Turn the vacuum on and start grooming him on his shoulder. The vacuum can be used to help keep a blanketed horse spotlessly clean, especially during nonbathing weather. It is also beneficial for lifting dirt out of a long winter coat.

Bathing the Horse

Bathing Equipment

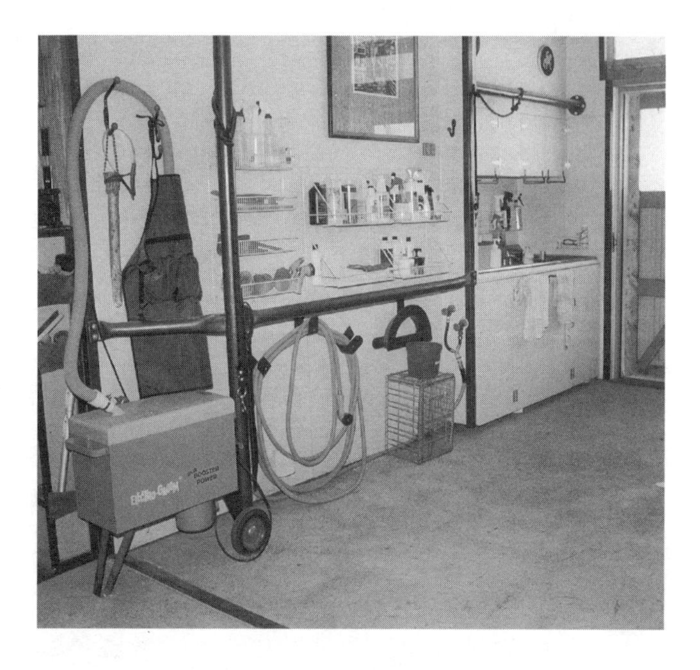

A Well-Designed Wash Area ▶

A well-designed wash area has a textured concrete floor that slopes toward a drain in the middle. The walls are waterproofed with vinyl paneling over plywood. The heavy metal posts and horizontal bars protect the items on the wall from the horse and the horse from the items on the wall.

Other features include:

- ✦ Hot- and cold-water taps
- ✦ Hoses and hangers
- ✦ Racks for storing shampoos and bathing supplies
- ✦ Large deep sink and long drain board for scrubbing buckets and brushes
- ✦ Cabinets above and below for storage
- ✦ Hooks and bars for towels and apron
- ✦ Cross ties at both ends
- ✦ Chest ropes/butt ropes at both ends

HERE ARE SOME of the items you will find useful for bathing. Top row: a sturdy step stool, plastic sweat scraper, aluminum sweat scraper, squeegee, combination shedding blade/sweat scraper, and bucket. Middle row: natural sponge, diluted shampoo, shampoo concentrate, medicated shampoo, conditioner. Bottom row: rubber mitt, rubber curries, hose brush, multijet sprayer, and small rinse cup.

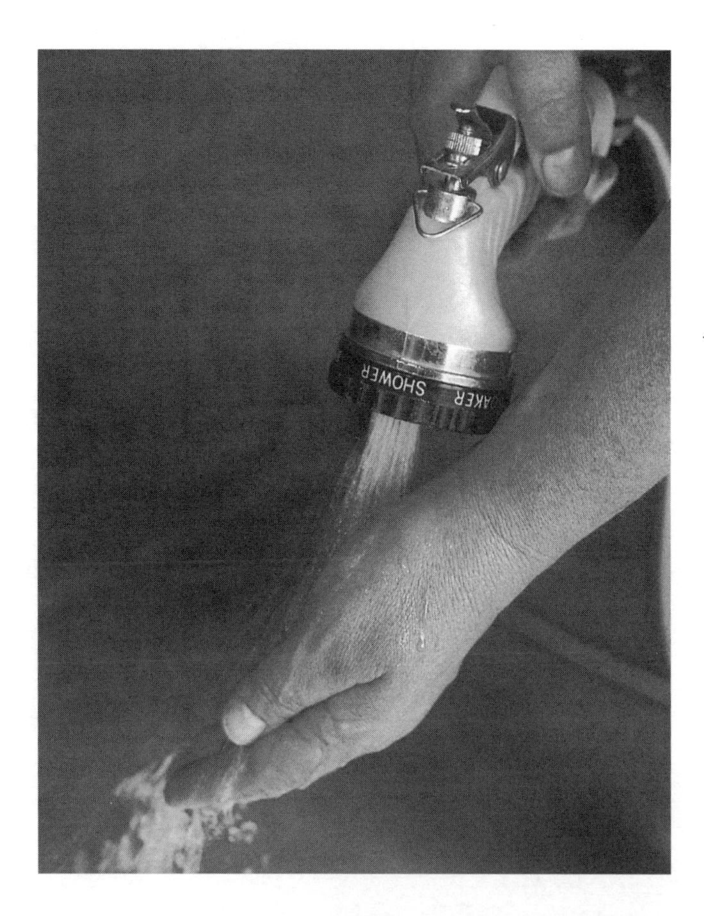

◀ WATER TEMPERATURE

Water that is comfortably warm to your hand makes the best temperature for bathing horses. It will do a better job of lifting scurf and debris from the coat than will cold water, and the horse will be more relaxed during the process. Cold water doesn't allow shampoo to suds and combine with the dirt to lift it away. Hot water could scald your horse's skin. Check regularly throughout the bath to be sure the temperature is remaining constant. If you don't have a hot-water tap available, set four or five 5-gallon pails of water out in the sun to warm for a few hours before you begin bathing your horse. Bathe your horse on a still, sunny day. Don't bathe when the temperature is below 55°F or when it is windy.

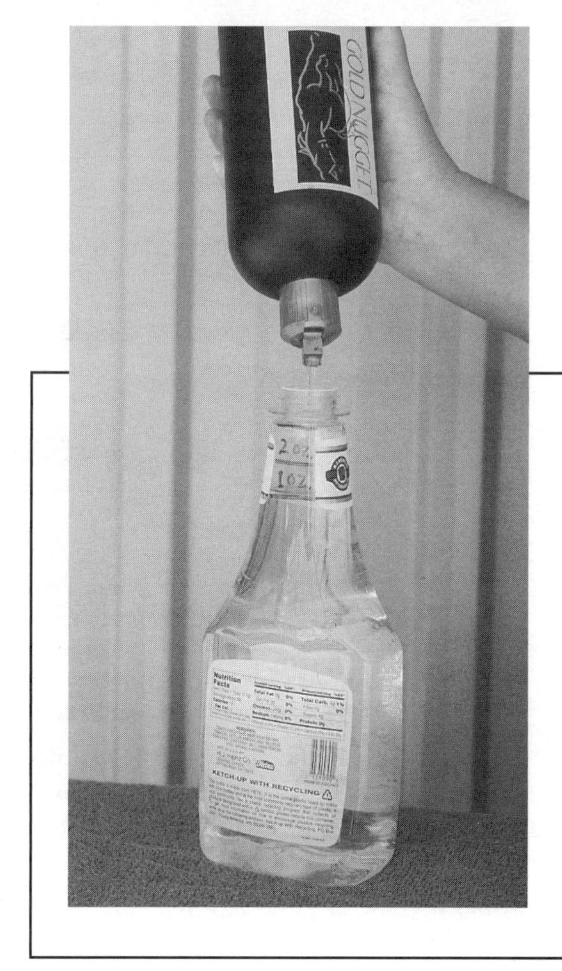

DILUTING SHAMPOO

To make dilute shampoo solution (9 parts water to 1 part shampoo), wash out an old squirt ketchup bottle or dish soap bottle and fill it close to 9/10 full of water. Squirt concentrated shampoo into the bottle of water, leaving a little bit of air space at the top. Close the squirt bottle and invert it several times (don't shake) to distribute the shampoo evenly in the solution. Using a waterproof marker, label the bottle with the name of the shampoo. Always use diluted shampoo solution on your horse; never use straight shampoo or it will take you much time and a great deal of water to rinse it out thoroughly.

Bathing the Coat

Shampooing

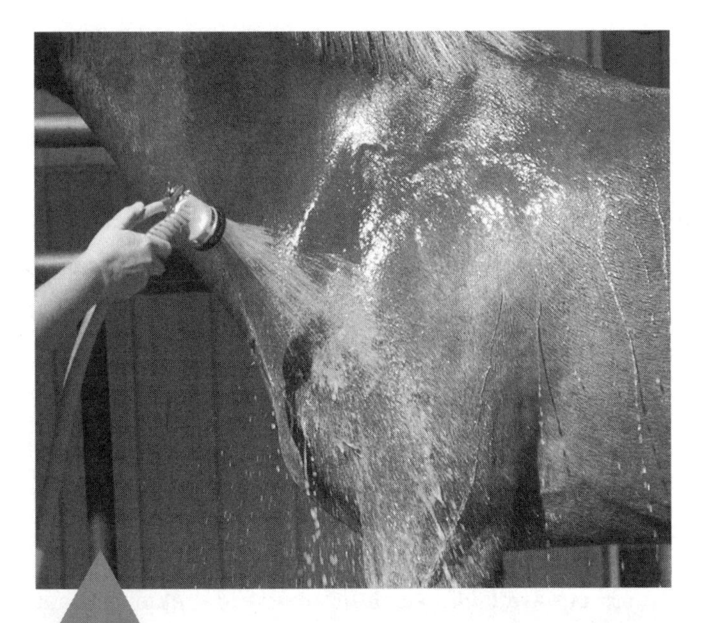

1 Start by wetting a section of the horse that has thick, deep muscles, such as the shoulder. Use a shower or mist spray rather than a forceful stream. Let your horse get accustomed to the idea that he is getting wet before you proceed.

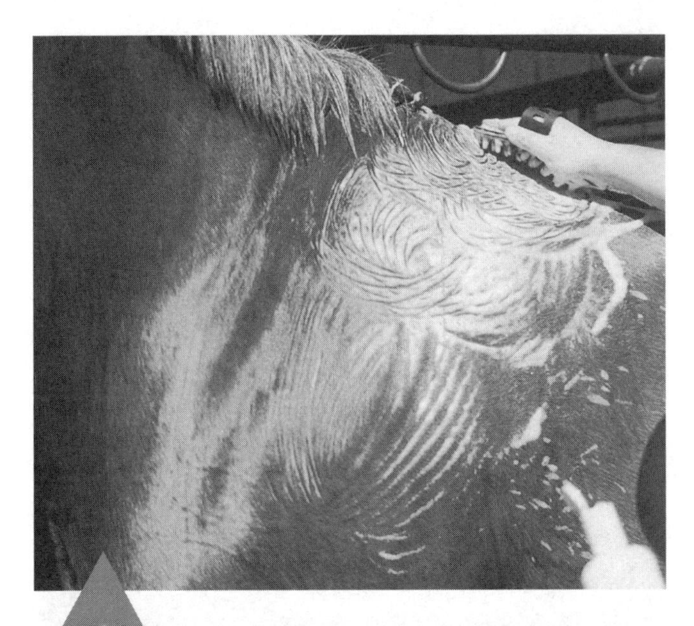

2 Squirt a little shampoo solution onto the wet portion and scrub with a rubber curry to lift debris from the skin and coat.

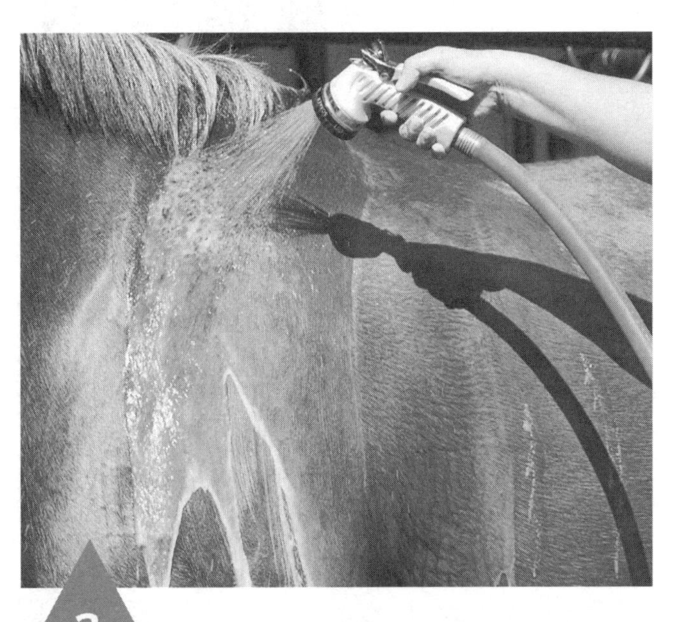

3 Rinse the coat thoroughly with a hose or using buckets of water. It is a good policy to rinse until you think you have all the shampoo removed and then rinse some more. The coat and skin should feel squeaky clean, not the slightest bit slippery. If you leave soap residue on the coat, the horse's skin could become dry and irritated and he could begin a bad habit of rubbing.

Removing Rinse Water

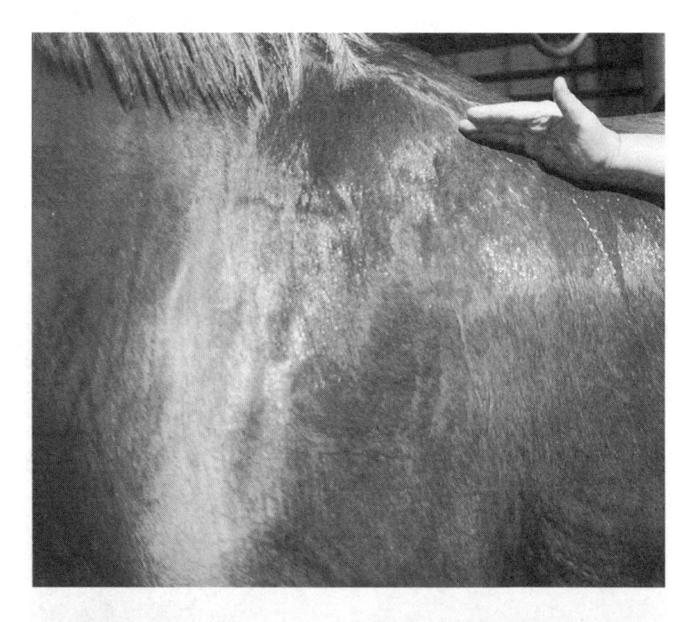

◀ **Using Your Hand**
One way to remove the last of the rinse water is with your hand, like this.

◀ **Using a Sweat Scraper**
Another way to remove the last of the rinse water is to use one of the sweat scrapers such as this combination shedding blade/sweat scraper. Note that the shedding blade teeth are facing outward for this use, while the smooth side is placed against the horse's body.

◀ **Completing the Job**
Follow the same procedure around the horse's entire body. I usually start at the near shoulder, then go to the ribs, back, and hindquarters on the near side. Then I move to the shoulder on the off side, the ribs, back, and hindquarters.

Washing the Head

After washing the horse's coat, you can wash the head. There are several ways of wetting and washing the head, depending on your horse's experience with bathing and how dirty his head and face are.

▲
Using a Sponge
Release a sponge full of water down the horse's face. Be sure that your horse is wearing a very clean, roomy halter.

▲
Using a Hose Sprayer
For the well-trained horse, the best way to wet the head is to use a hose sprayer on the light mist setting. A hand at the bridge of the horse's nose reminds the horse "head down."

Using a Wet Towel

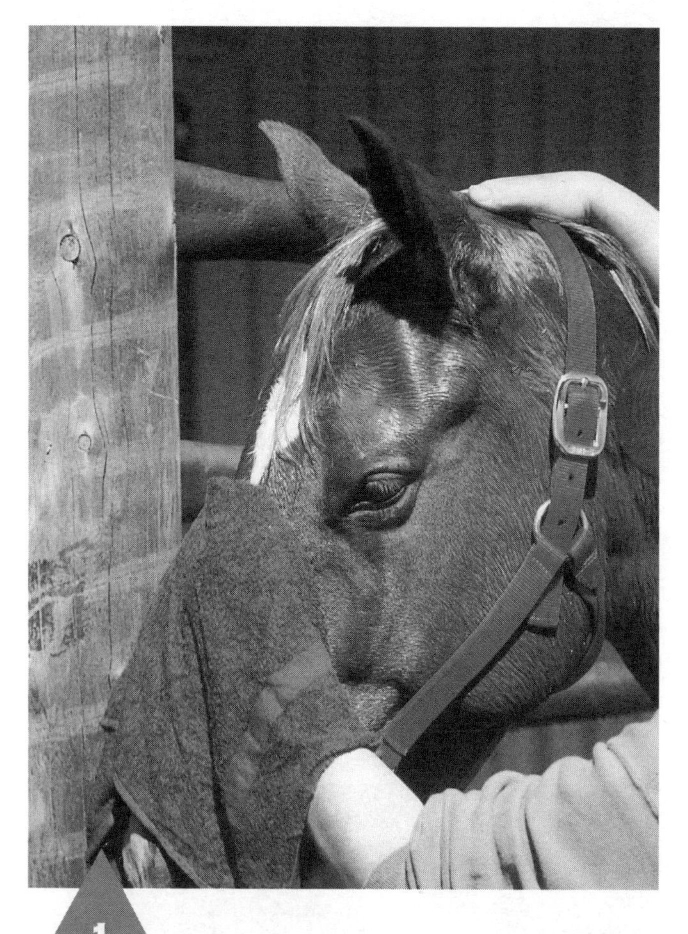

1 For the young horse's first bath or for a horse who isn't dirty enough to require a thorough head shampoo, you can wet the head with a damp towel. The right hand at the poll reminds the horse "head down." You can use one cloth with a tiny amount of shampoo solution on it and follow it with a cloth wet in clear water. Rinse the last cloth several times as you remove dirt and shampoo. Continue wiping with a clean, wet towel until the towel rinses clear.

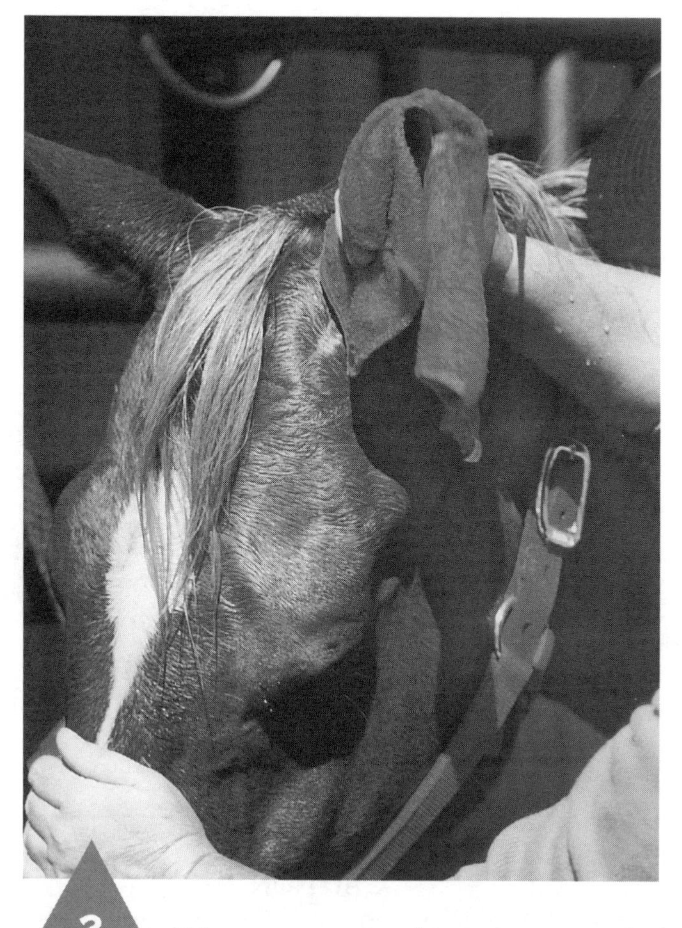

2 Take care not to drip water into the horse's ear as you rub it firmly from base to tip, inside and out.

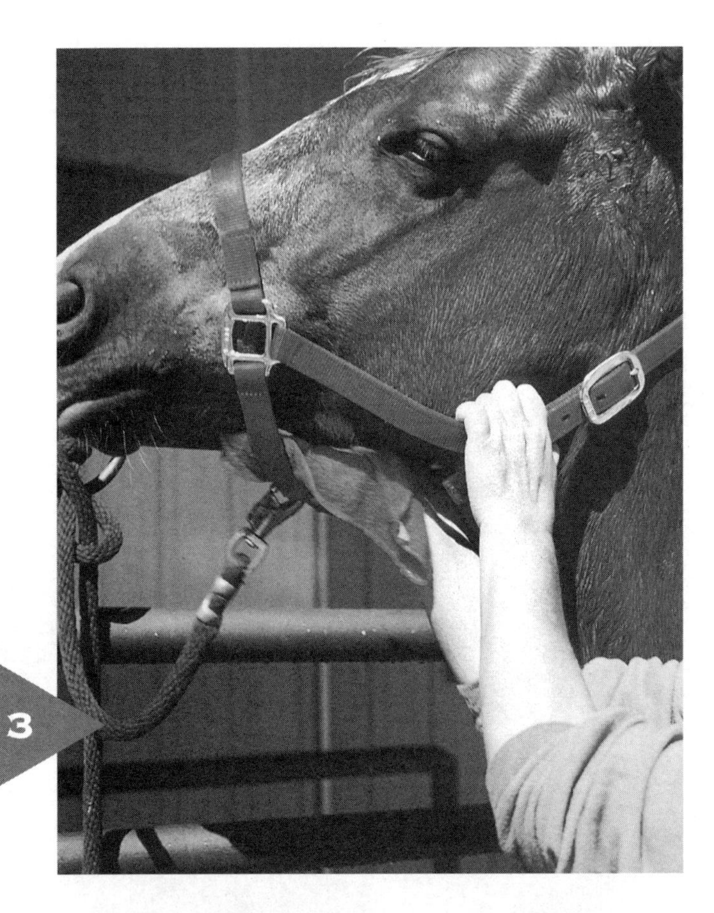

When cleaning the underside of the jaw and throat, it is a natural reflex for the horse to raise his head. That actually makes it easier for you to do a good job. Note the benefit of using a large, roomy halter for head bathing. It gives you room to work.

CAUTION

Don't use shampoo near the eye, nostril, or mouth area. Just use a wet cloth and wipe out the nostrils thoroughly. Keep your hand on the lead rope to remind the horse to keep his head low if the procedure starts to annoy him.

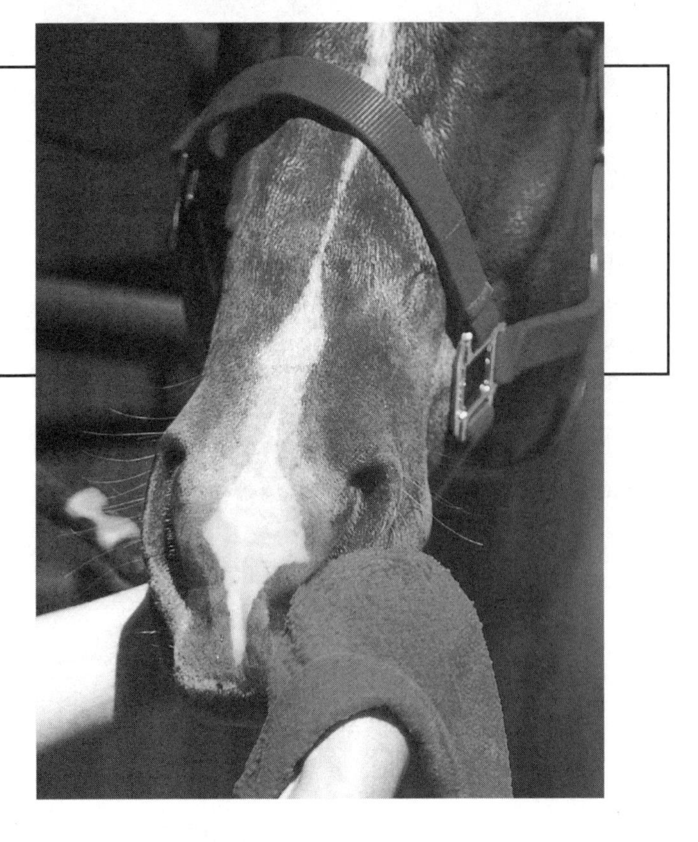

Shampooing the Mane

Most often the mane is shampooed and rinsed at the same time as the neck.

1 Wet the mane down the crest line. You may have to stand on your stool to reach. A hand on the bridge of the horse's nose reminds him "head down."

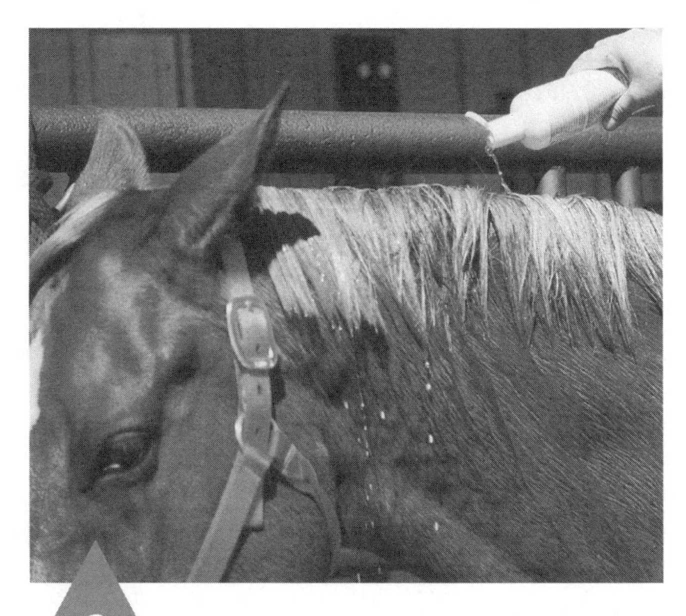

2 Dribble some shampoo solution along the crest, but not too much or you will spend hours rinsing!

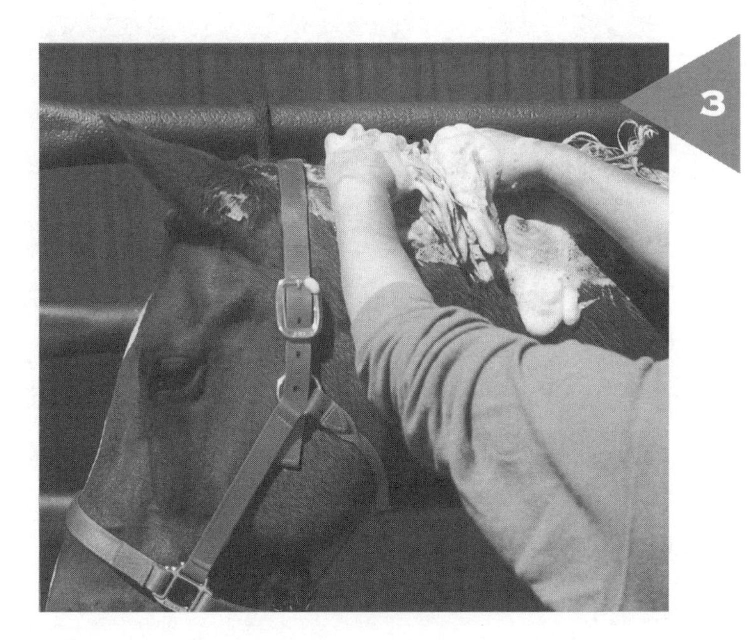

3 Work up a lather by rubbing vigorously from poll to withers and back again several times.

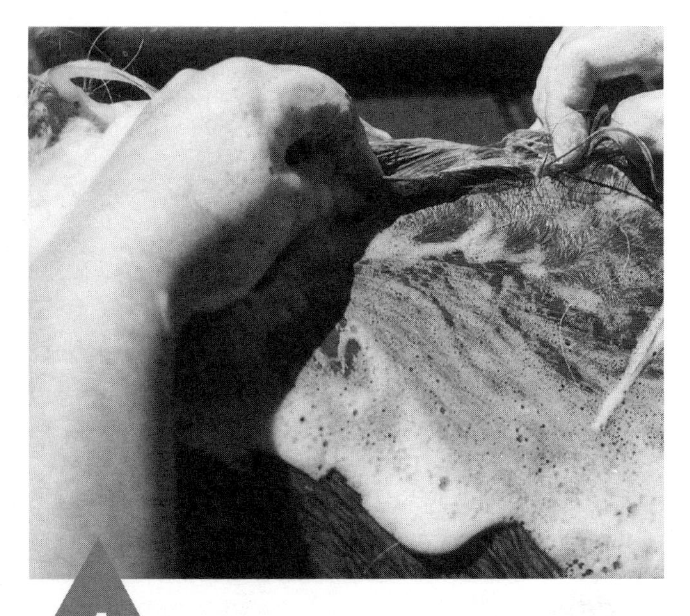

4 Get down to the roots of the mane hairs and the skin of the crest where scurf and dirt will most likely be hiding in layers.

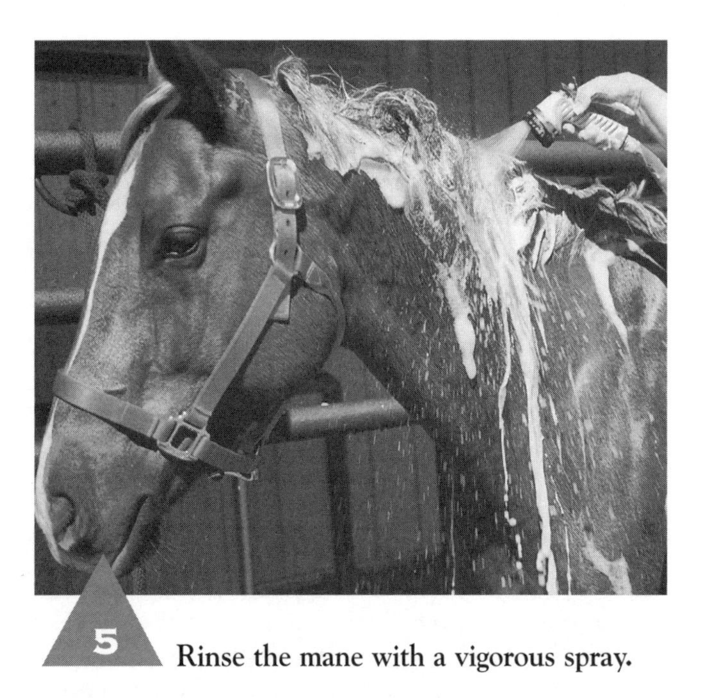

5 Rinse the mane with a vigorous spray.

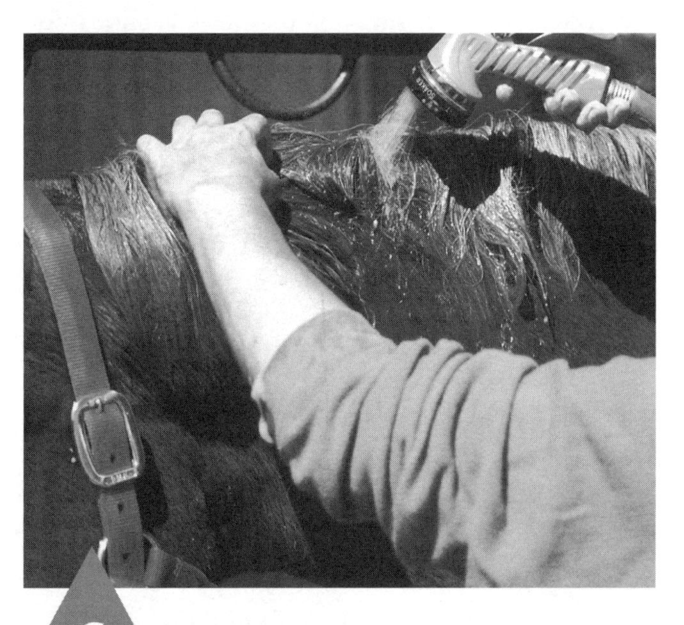

6 After all visible signs of shampoo are gone, continue rinsing for another minute or so.

SHAMPOOING THE TAIL

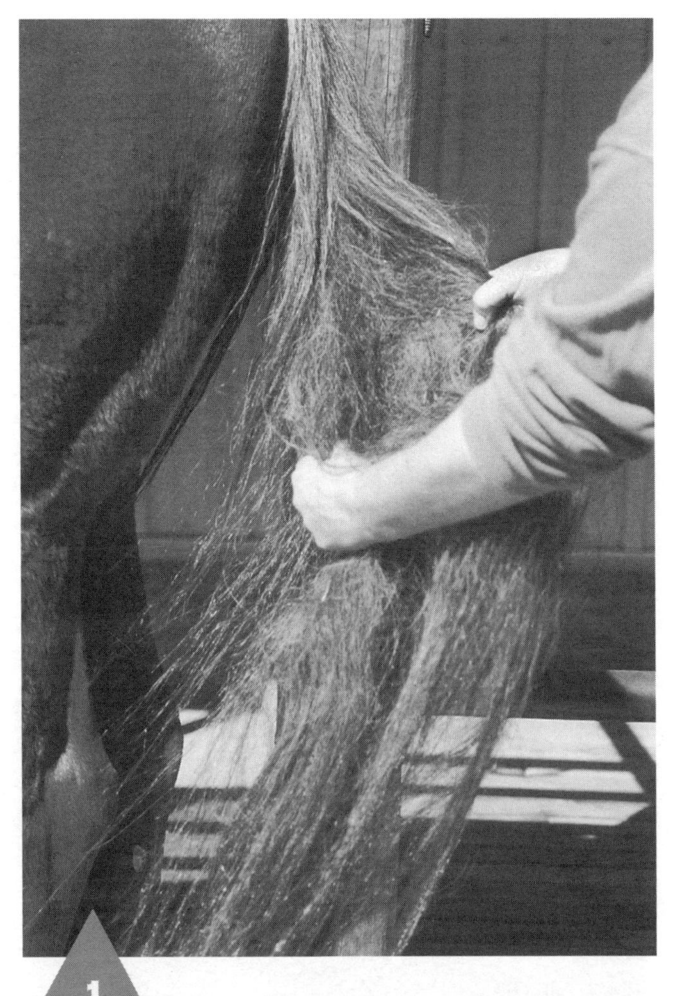

1 Before shampooing the tail, you must get rid of all the tangles in it. This tail has been "turned out to pasture" for the winter so it needs a lot of fingering through before shampooing. If you try to shampoo this matted tail, you will end up with a wet, matted, much more difficult to separate, mess — and the horse will end up losing much of his tail hair. There are several good detangling products on the market that will help you remove knots and burrs.

2 With most tails that are regularly cared for, all you need to do is lightly finger through the tail hairs, removing any hay or tangles.

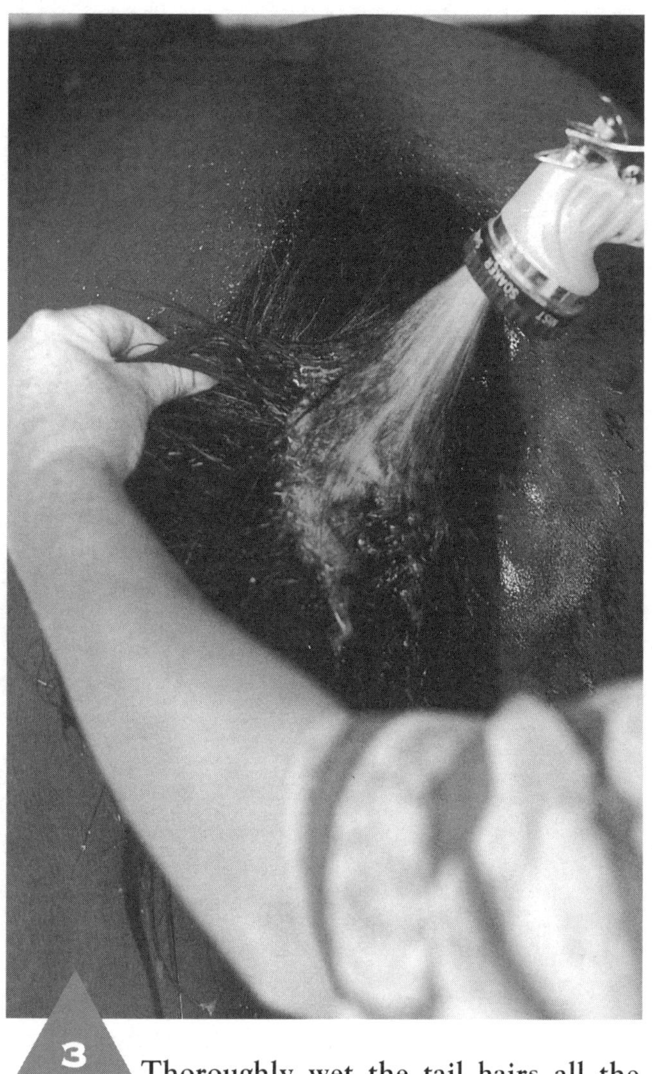

3 Thoroughly wet the tail hairs all the way down the dock (the fleshy covering of the tail bones) and including the skirt of the tail all the way to the ground.

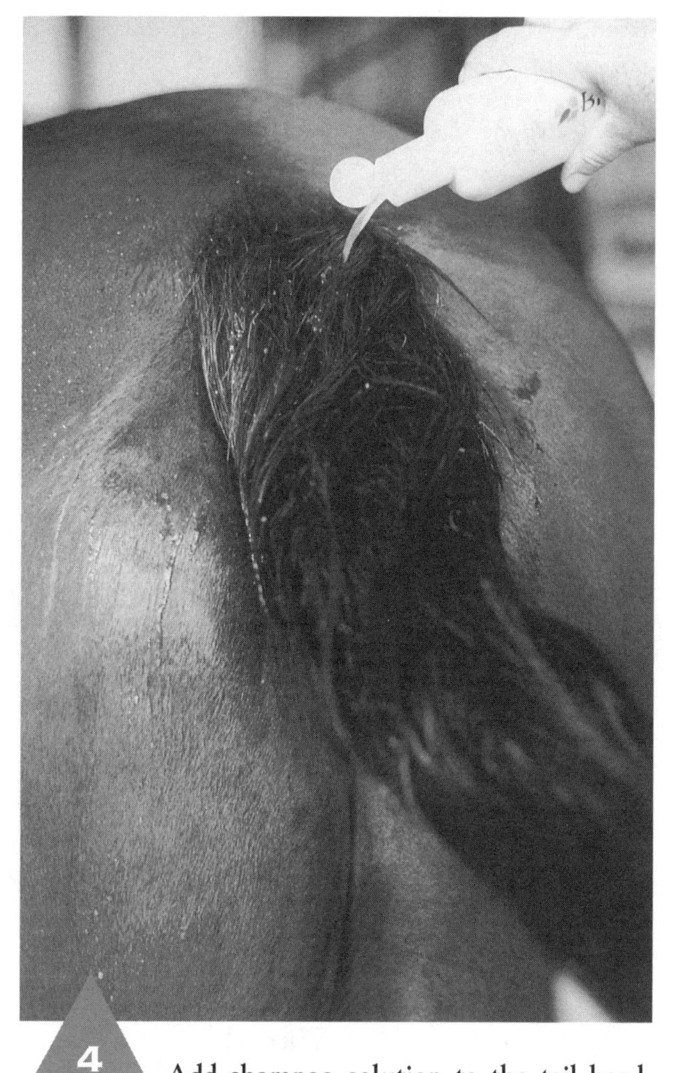

4 Add shampoo solution to the tail head. The main portion you need to shampoo is the dock.

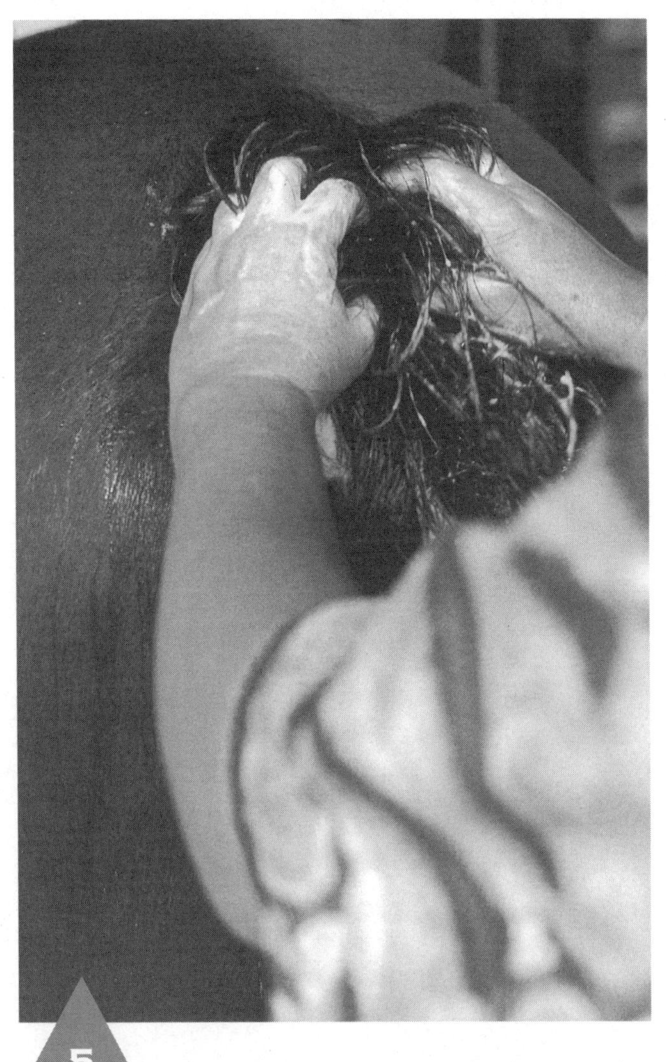

5 Scrub vigorously all the way down to the skin. Your horse will enjoy this immensely because the tail head is not a spot a horse can reach easily when self-grooming.

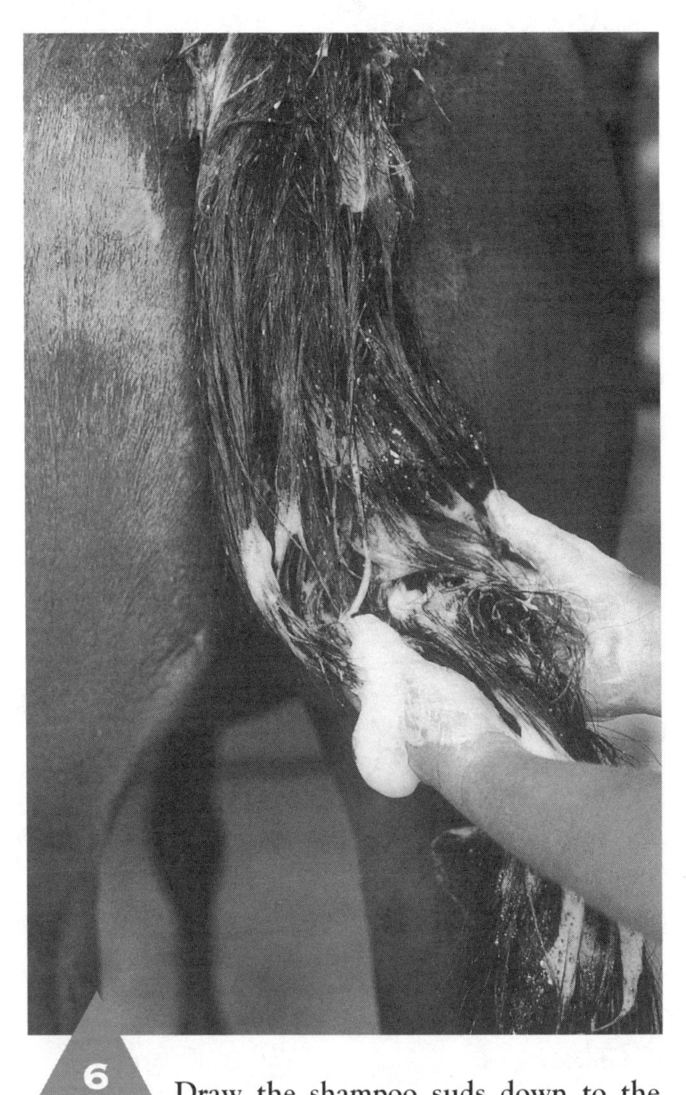

6 Draw the shampoo suds down to the skirt of the tail, working it through with a scrubbing action. Take care not to tangle the long hairs unnecessarily.

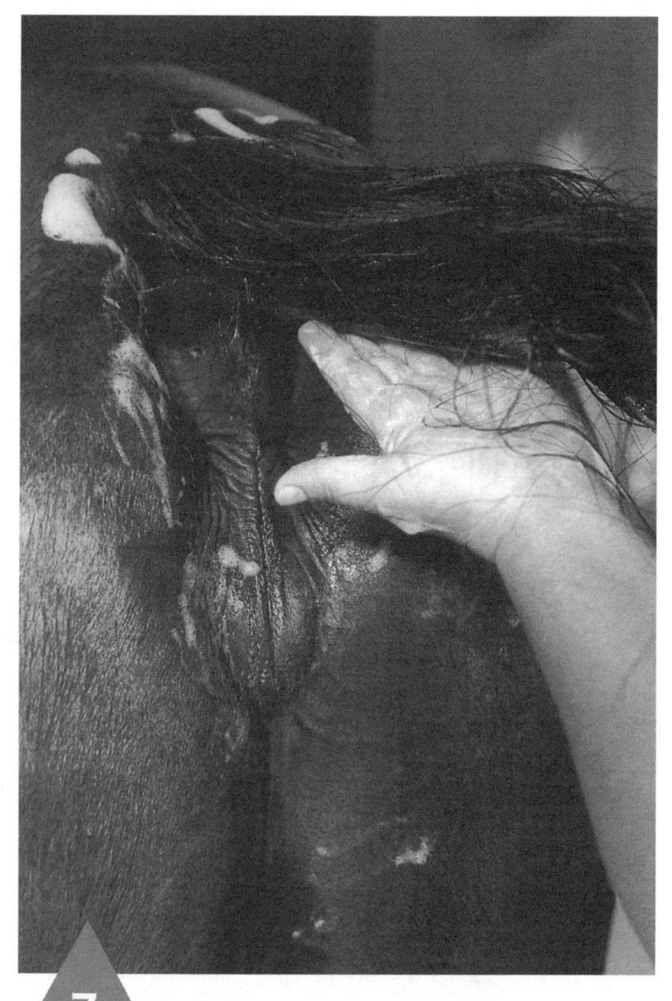

7 Wash the underside of the dock where the smooth skin lies against the rectum (and vulva in females).

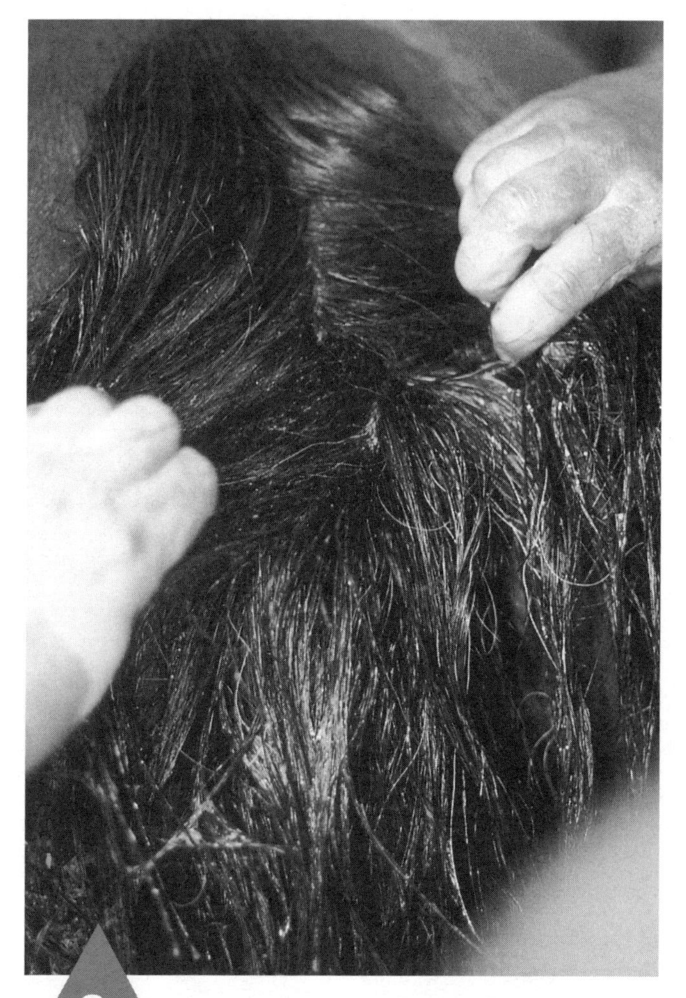

8 Separate the hairs at the tail head so you can see if you removed all the skin flakes and dirt.

9 Begin rinsing and do an even more thorough job than you have done on the body and mane. Why? If you leave soap residue on the dock, it will probably cause an uncomfortable itching sensation. The result can be tail rubbing, a hard habit to break once it is established. It can destroy a horse's tail and be hard on fences as well. Prevent this bad habit by keeping your horse's tail carefully maintained.

10 After shampooing, always use conditioner on your horse's tail. Read the label to see if the conditioner is designed to be left in or rinsed out. Dilute the conditioner just as you did the shampoo. Pour it over the tail head region and work it in with your hands.

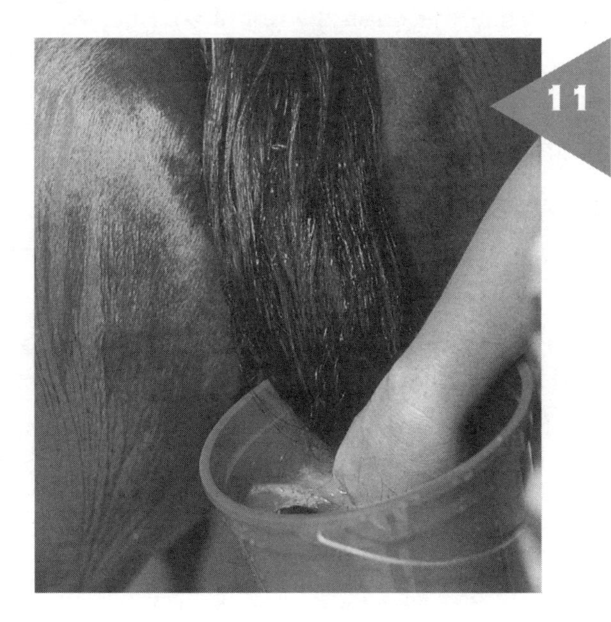

11 A good way to condition the skirt of the tail is to dunk it into a bucket of conditioner solution. If the conditioner is designed to be rinsed out, do so.

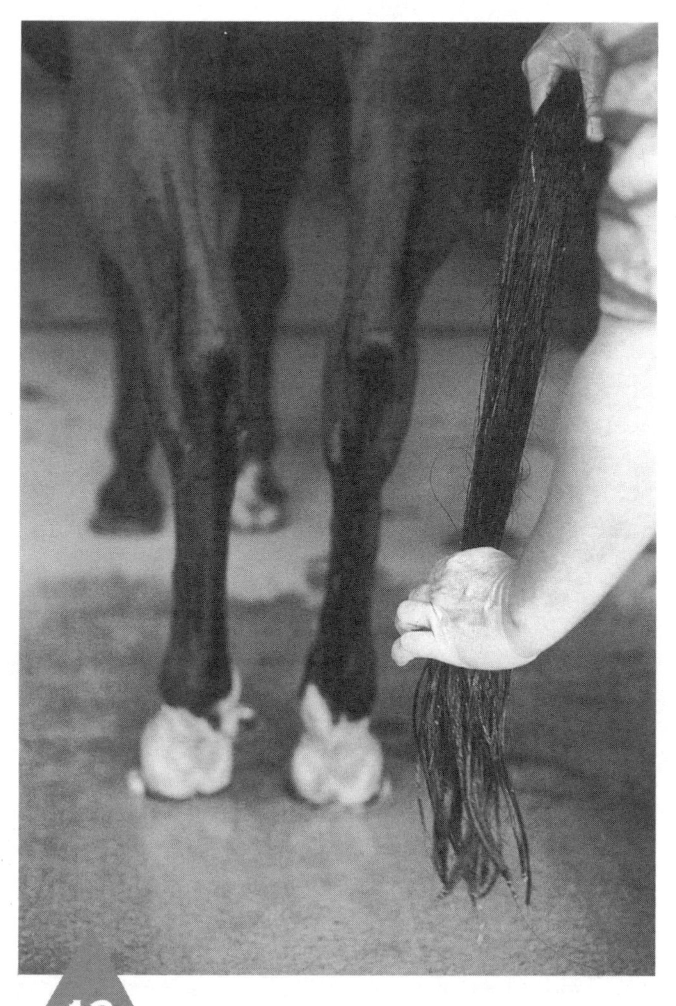

12 Remove most of the water from the tail by stripping it with your hand.

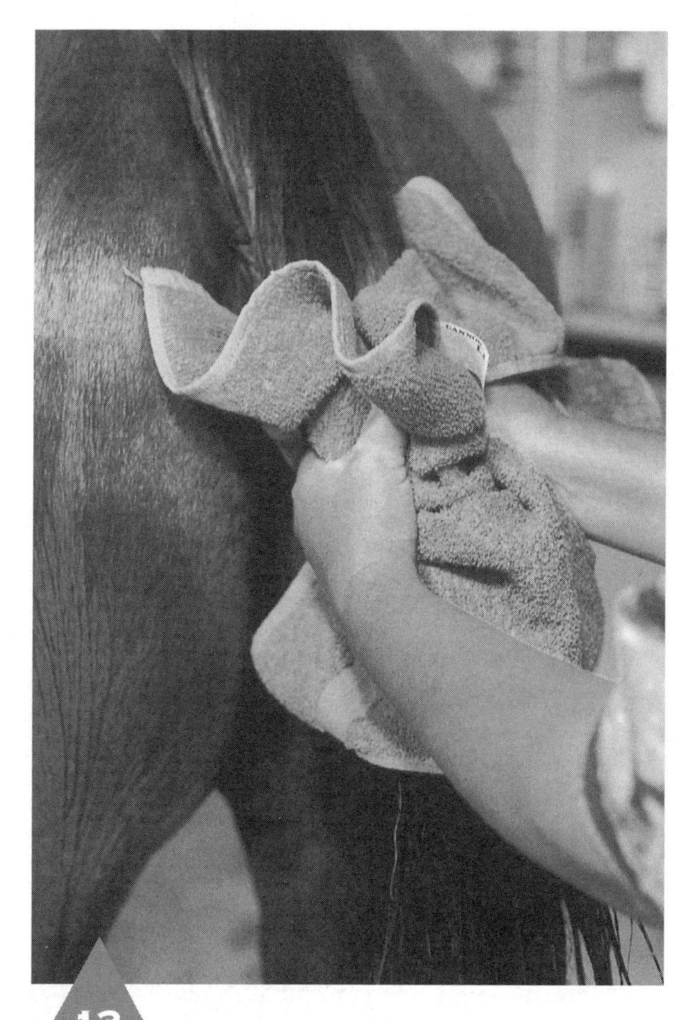

13 Dry the tail further by squeezing it with a dry terrycloth towel. You may need to use several. The more water you remove at this stage, the faster the tail will dry and the sooner you will be able to finish it. For the time being, however, leave the tail while you finish the rest of your horse.

WASHING THE LEGS

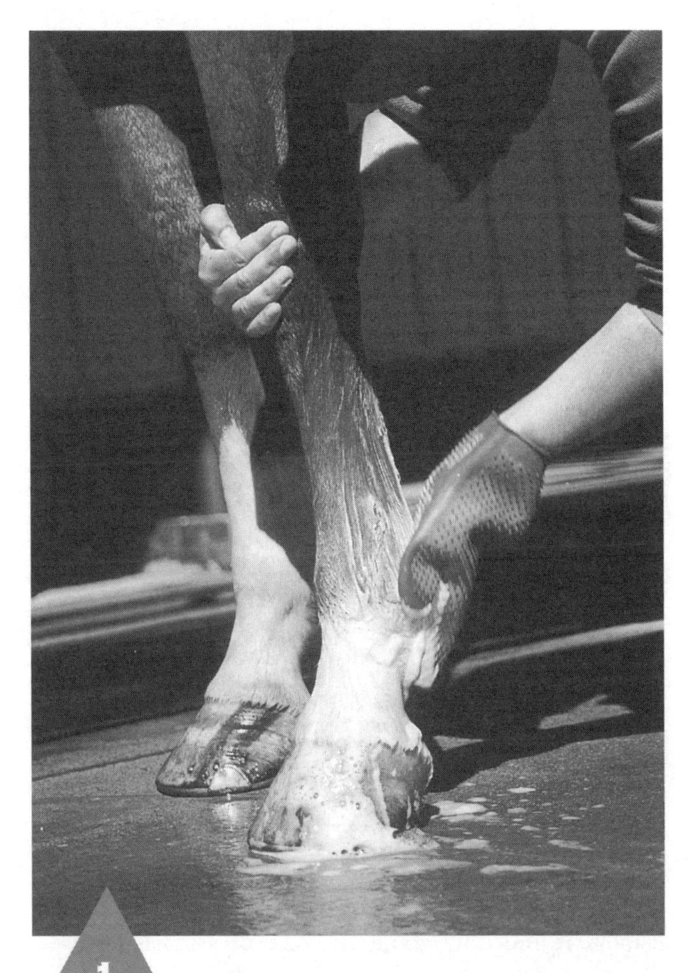

1 During all the body, mane, and tail washing, your horse's legs have probably gotten pretty wet, but if they have dried off, wet them again. Using the dilute shampoo solution, scrub the horse's legs with a rubber mitt. To let your horse know that you don't want him to pick up his leg, use your other hand to press back on his knee. This locks his knee and is a signal for him to keep his leg straight, not to bend it. At the same time, use a voice command such as "Stand" or "Stand on it."

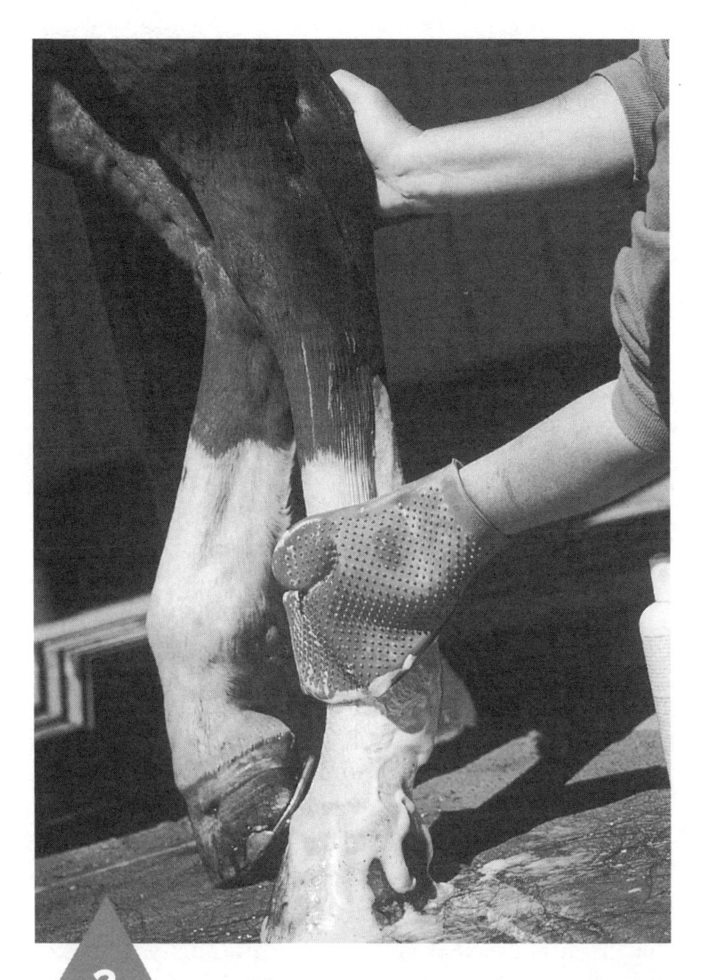

2 Similarly with the hind leg, if you push forward on the hock, this will tend to keep the hock locked and the horse's weight on the leg you are working on. After you have scrubbed the legs, rinse them well. When you have finished bathing your horse's legs, remove chestnuts and ergots.

Removing Chestnuts and Ergots

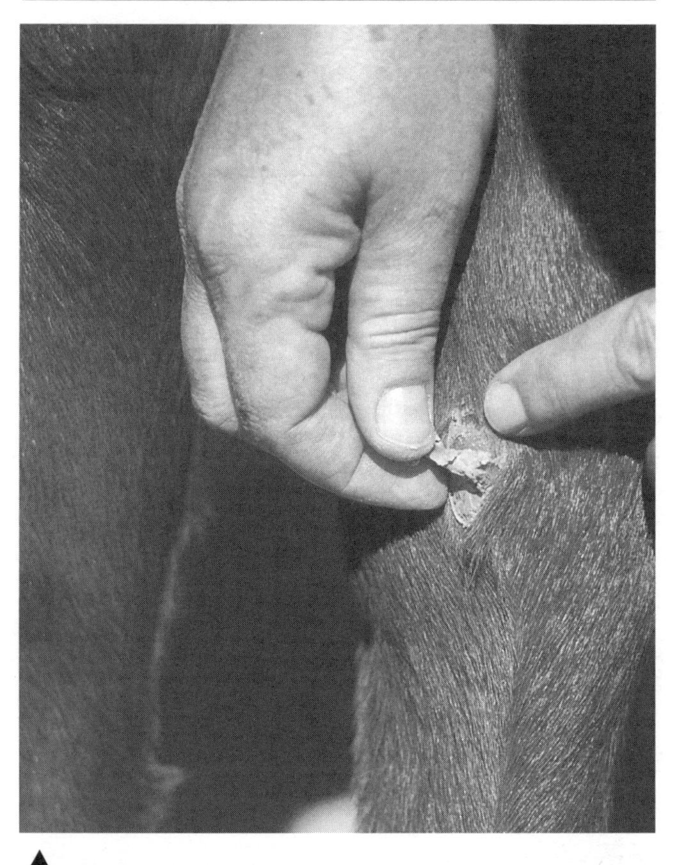

▲
Chestnuts
Chestnuts are horny, insensitive growths on the legs that need to be kept short and neat. This will make them less likely to get ripped off and make it easier for you to clip your horse's leg hair. On the front legs, chestnuts appear on the inside above the knee. On the hind legs they are on the inside just below the hock. On some horses chestnuts are barely noticeable, on others they can be quite large and thick. To keep chestnuts close to level with a horse's coat, peel off softened layers after a bath or trim hardened layers with sharpened nippers.

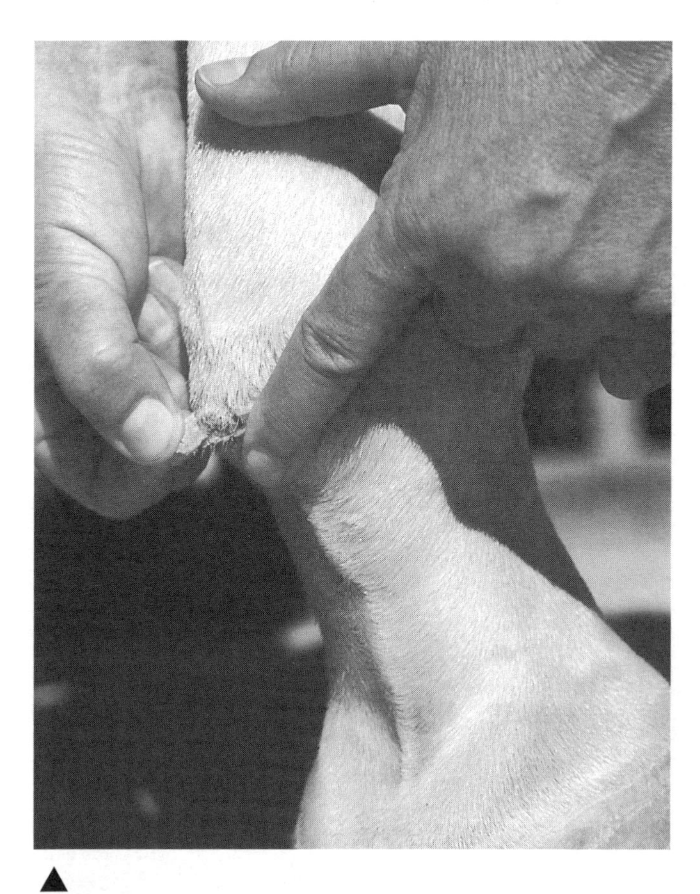

▲
Ergots
Ergots are horny stubs similar to chestnuts. They are usually round dots about the size of a small button. Ergots appear at the rear point of the fetlock on all four legs. As with chestnuts, some horses have heavy, large ergots and others have almost no ergots at all. The best time to peel them is after a bath when they are soft.

Difficult-to-Remove Ergots ▶
If an ergot is particularly dry or difficult to peel, loosen the edges and twist it off the center stalk. Hold the horse's leg when removing an ergot to prevent pulling the flesh around it, which could cause some discomfort to your horse. As with chestnuts, if ergots are very long, you may have to remove them with sharpened nippers.

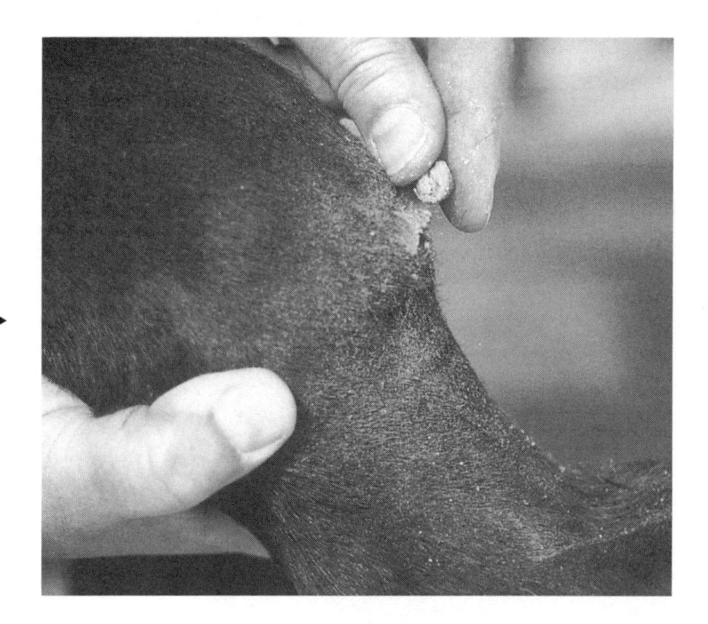

Finishing the Bath

Now it's time to condition the coat. (This is optional.) If your horse has a dry coat or if you have been using a medicated shampoo to deal with a skin problem, you might want to condition his entire coat to restore some oils to the skin.

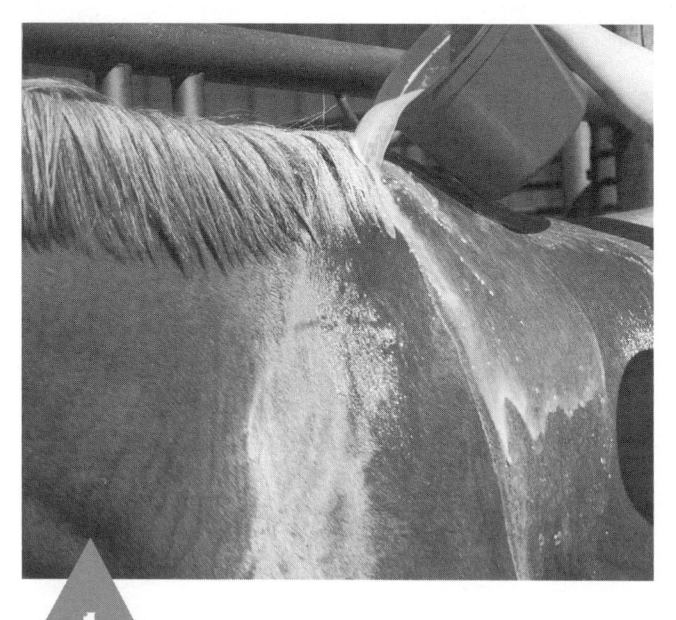

1 Pour the diluted conditioner solution all over his body and work it in with your hands. Let it set for a few minutes.

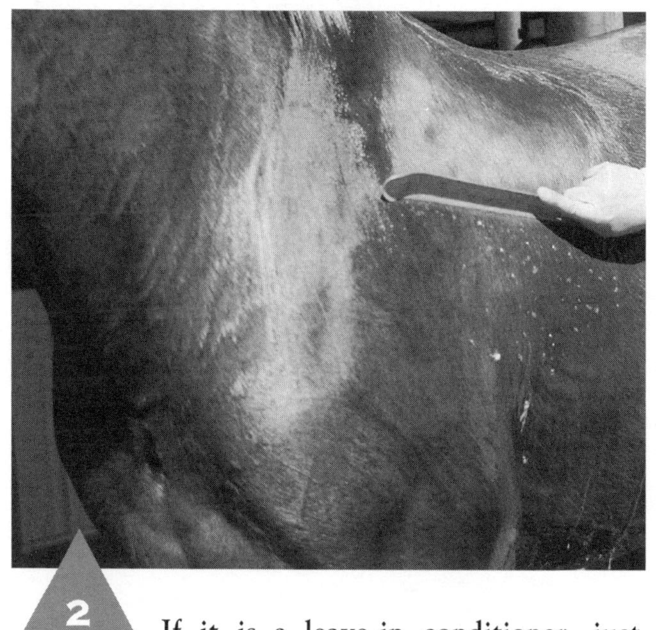

2 If it is a leave-in conditioner, just remove the excess with a sweat scraper. Note the particles of conditioner that did not fully dissolve in the dilute solution.

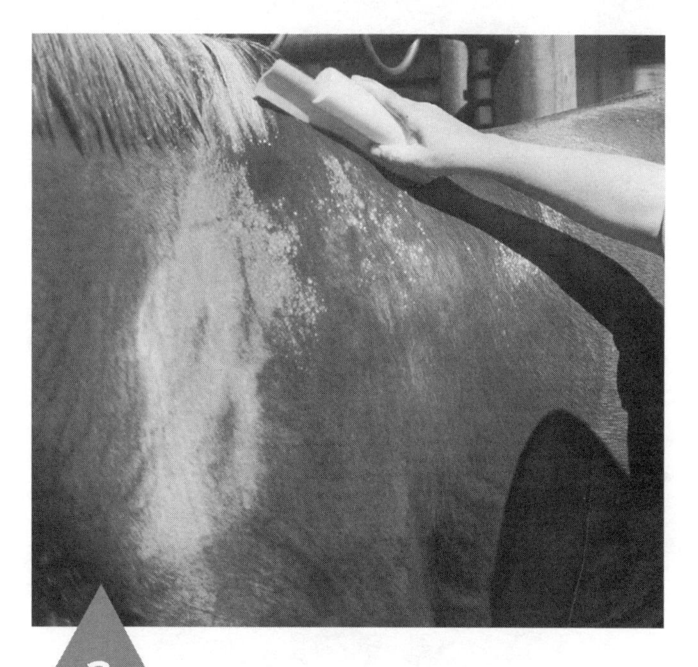

3 If it is rinse-off conditioner, rinse and then squeegee the excess water off the horse.

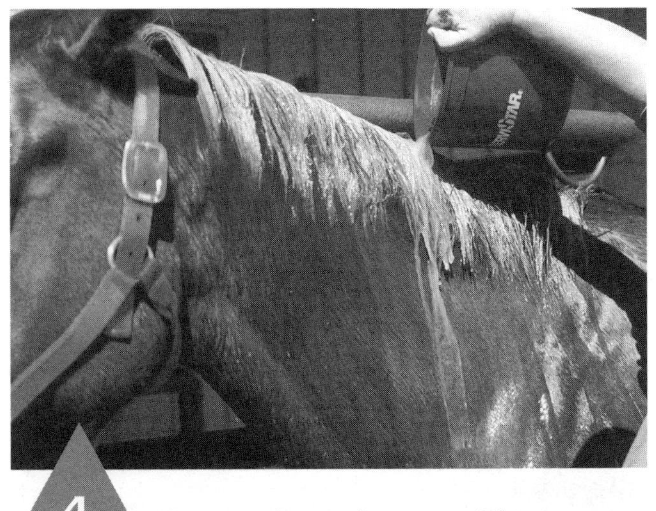

4 Now condition the mane. This is essential. Save some of the diluted conditioner and apply it by hand to the forelock. Leave in or rinse out as directed by the product label.

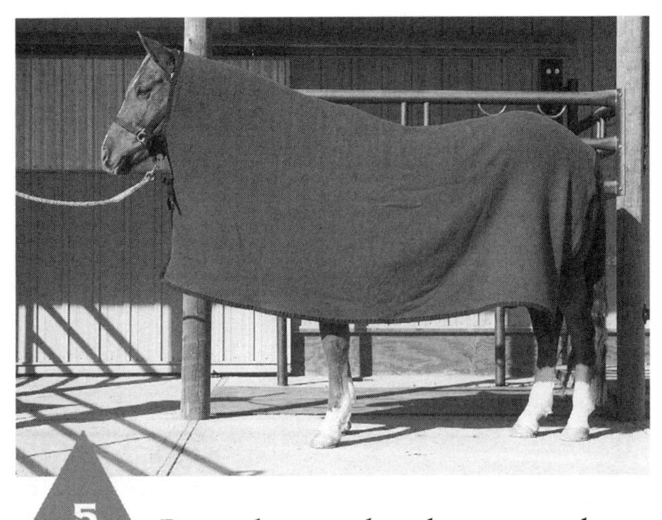

5 Put a clean wool cooler on your horse and let him stand in a warm place free of drafts. A wool cooler is a loose-fitting blanket that is draped over the horse and secured by a tail loop and front fasteners. If you think your horse will roll (which 99 percent of horses want to do after a bath!), you will have to tie him somewhere while he dries.

THE FRONT OF the wool cooler fastens with a brow band and two sets of ties: One set attaches to the halter side rings and one set ties at the horse's chest.

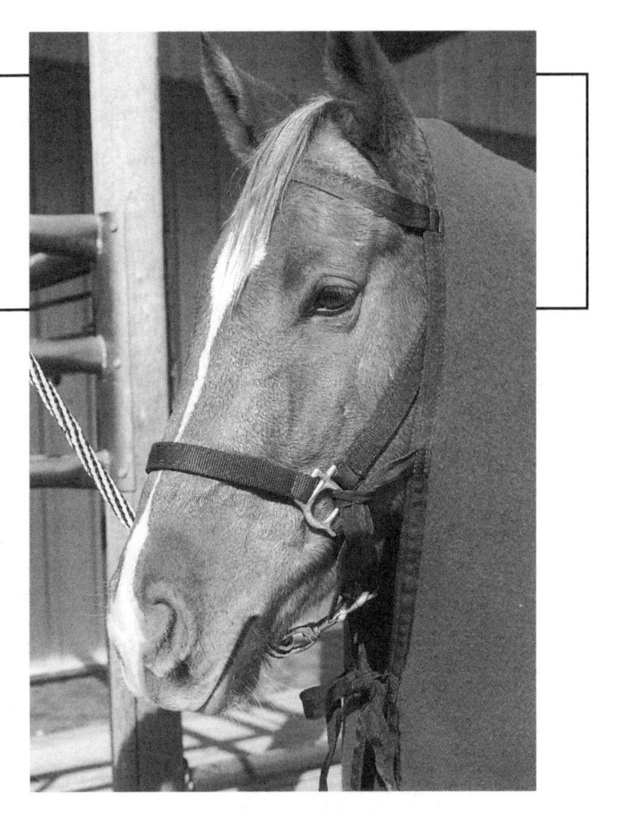

Conditioning the Tail

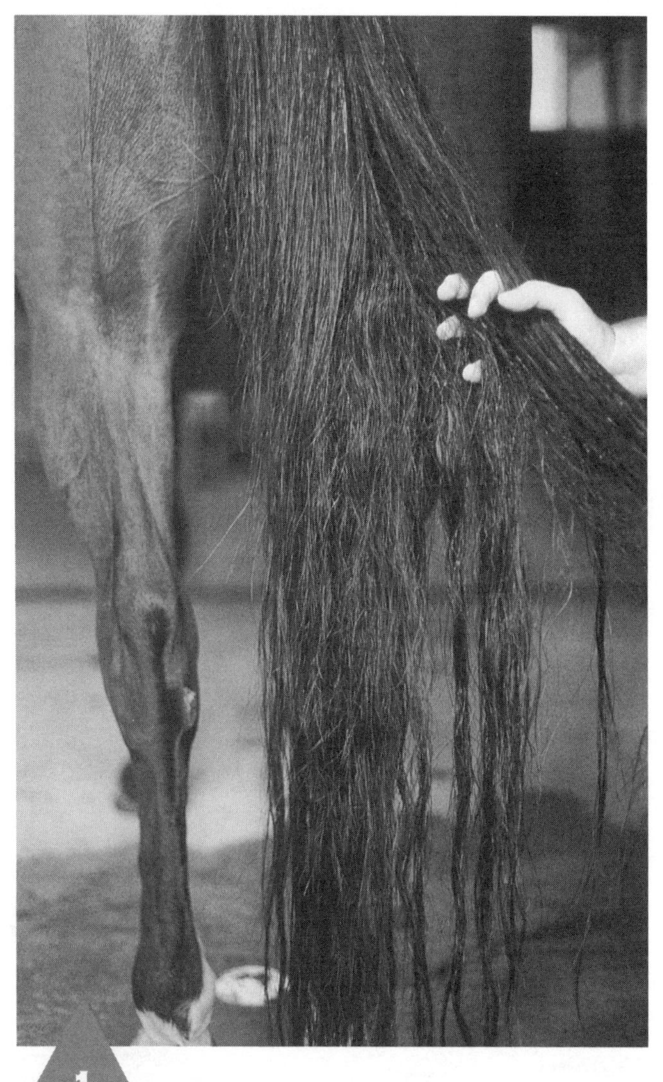

1 When the tail is almost dry, you can begin fingering through it.

2 As you work through the tail, you can spray on a leave-in conditioner, which often makes the very long hairs more cooperative.

3 When you begin brushing, use a human's hairbrush with bristles set far apart. Start at the bottom of the tail and work your way up.

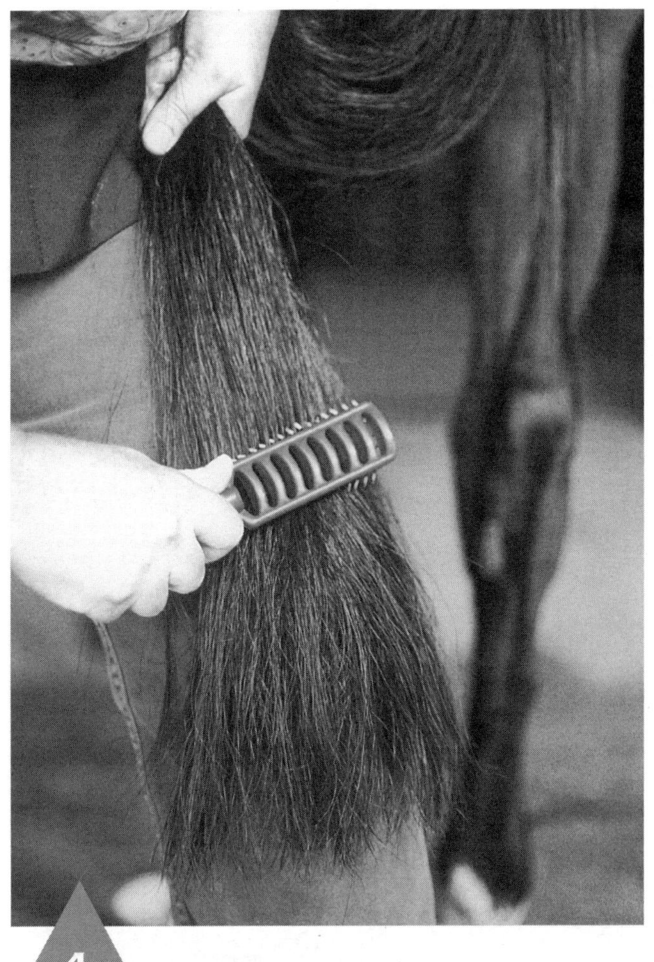

4 Rather than bending down to the floor, you can bring a long tail onto your leg to begin brushing.

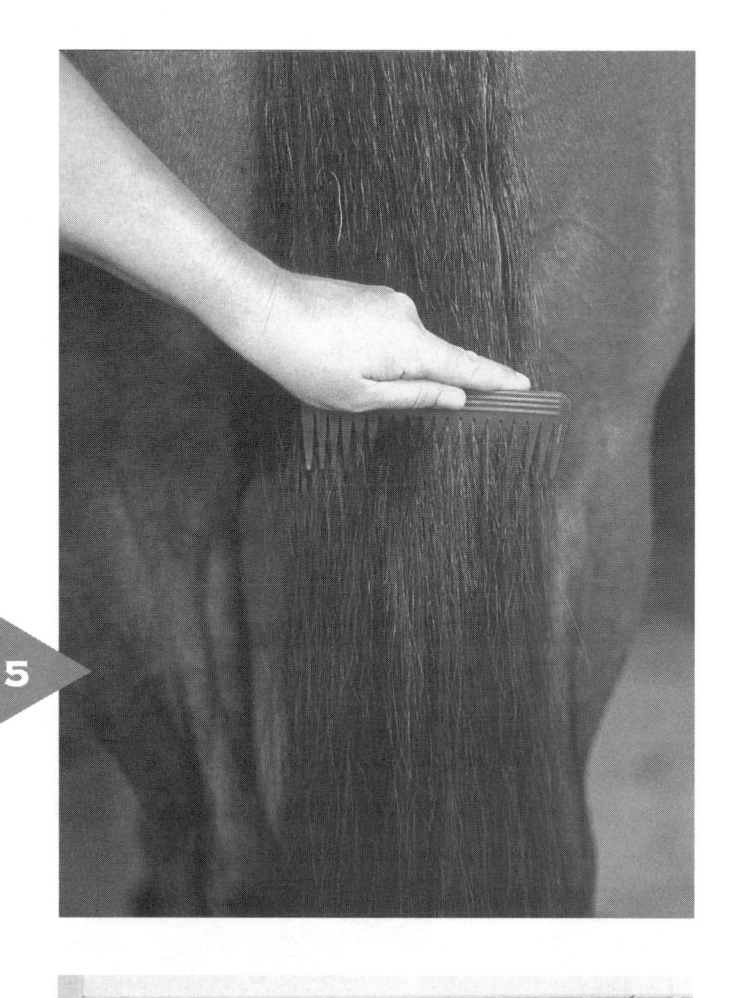

5

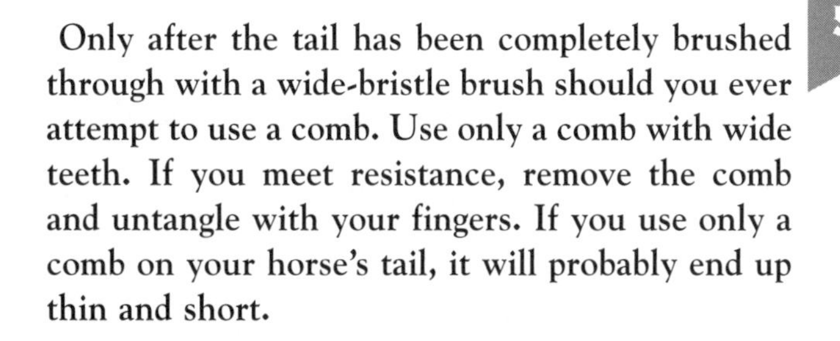

Only after the tail has been completely brushed through with a wide-bristle brush should you ever attempt to use a comb. Use only a comb with wide teeth. If you meet resistance, remove the comb and untangle with your fingers. If you use only a comb on your horse's tail, it will probably end up thin and short.

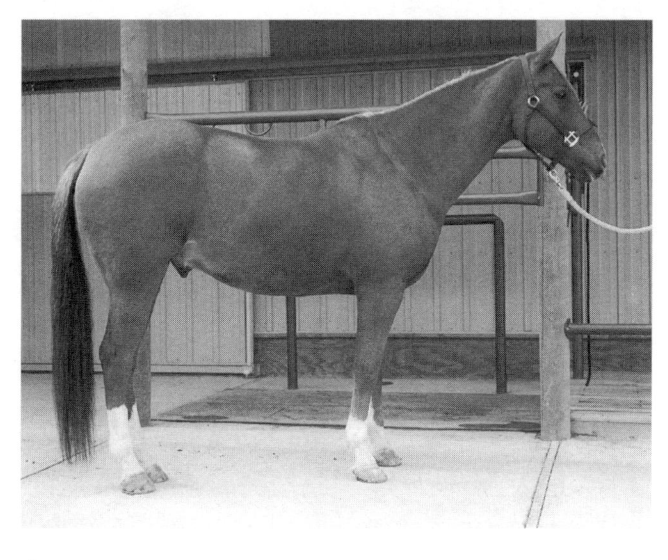

▲

The Clean Coat

When the horse is completely dry, although he is clean and he looks clean, he does not look dazzling. That's because freshly shampooed hair tends to stand up and "stare."

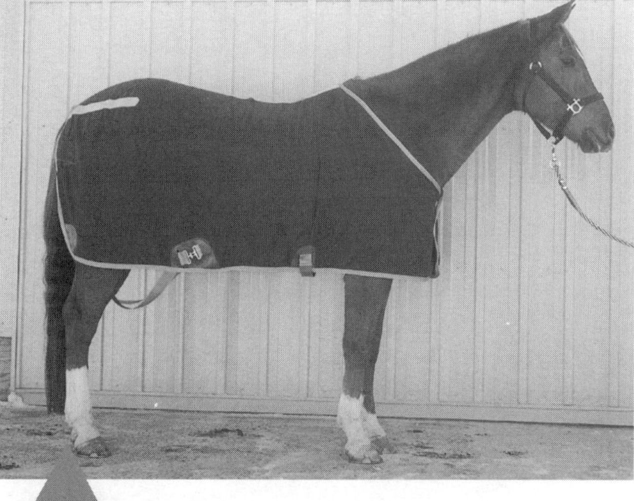

6 To lay the coat down, put a clean, well-fitted sheet on your horse. It will take about one day for his hair coat to lay down and for the skin oils to restore the gleam to the hair. Therefore, if you are planning a photo session or are preparing your horse for a show, give him his bath at least one day before the scheduled event and keep a clean sheet on him until then.

Clipping the Horse's Hair

To Clip or Not to Clip

There are certain portions of your horse's body that are covered with extra hair, for a reason. Be sure you understand the purpose of the hair on your horse's body before you remove it. Often show horses are clipped just about everywhere and are kept very short-coated. That's because they live in a barn and are protected by blankets and hoods. If your horse lives au naturel, your clipping routines should be designed to tidy him up but not drastically alter his protective hair.

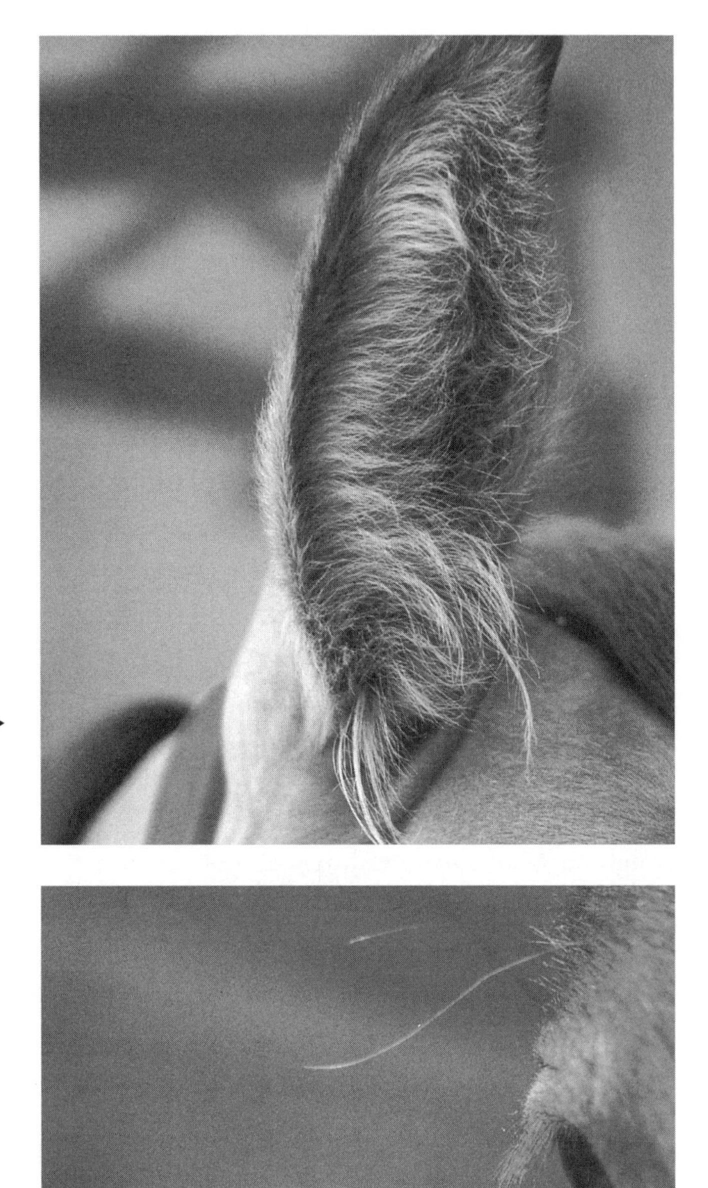

Hair That Lines the Ears ▶
Besides keeping the horse's ears warm in cold weather, the fine hair that lines them keeps out flies and even pesky gnats.

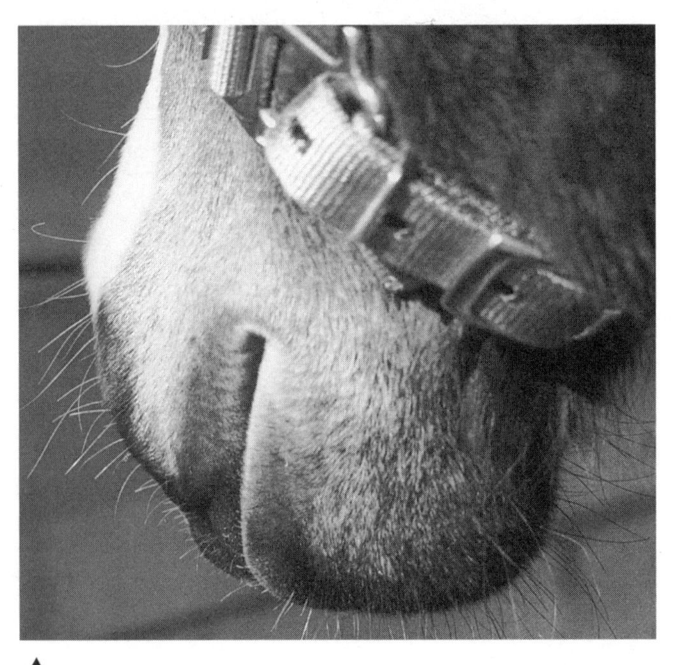

▲
Muzzle Hair
Muzzle hairs are like antennae for the horse's nose. When he puts his lips into a bucket or a feeder, he cannot see where he is sticking his nose. These "feelers" let him know where the sides of the bucket are. Without muzzle hairs, a horse could more likely injure his head.

▲
Hair around a Horse's Eyes
The hairs around a horse's eyes are also there for a reason. Horses' eyes are located in a vulnerable spot on the side of the head. These hairs are feelers that act like an early warning system.

PREPARATION FOR CLIPPING

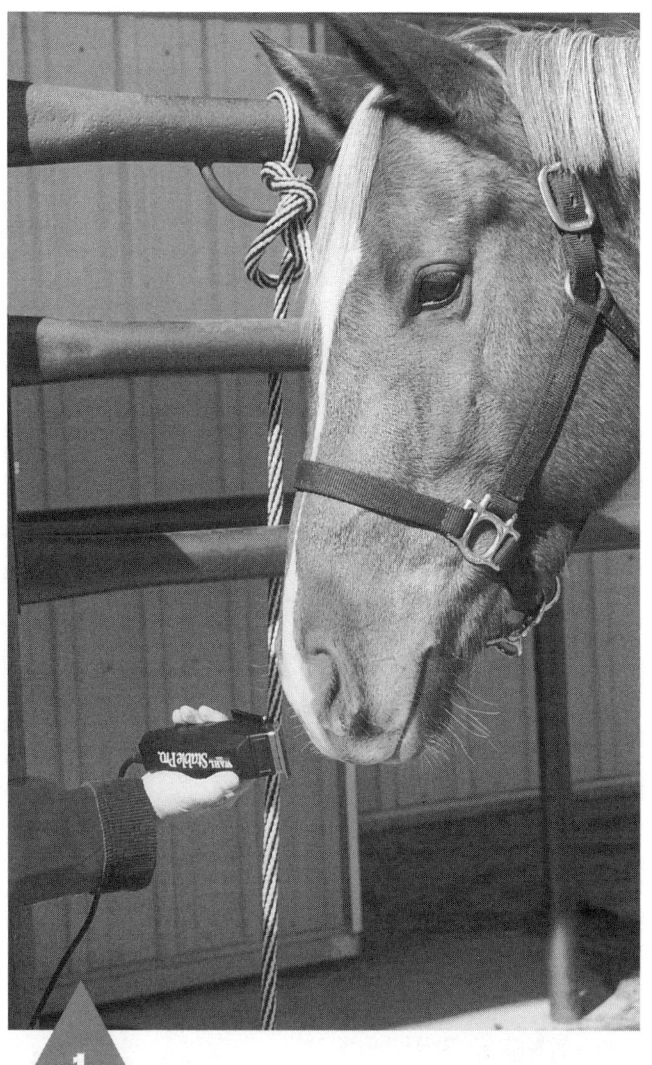

1 Let your horse look at and smell the clippers. Turn them on and let him hear the sound that they make.

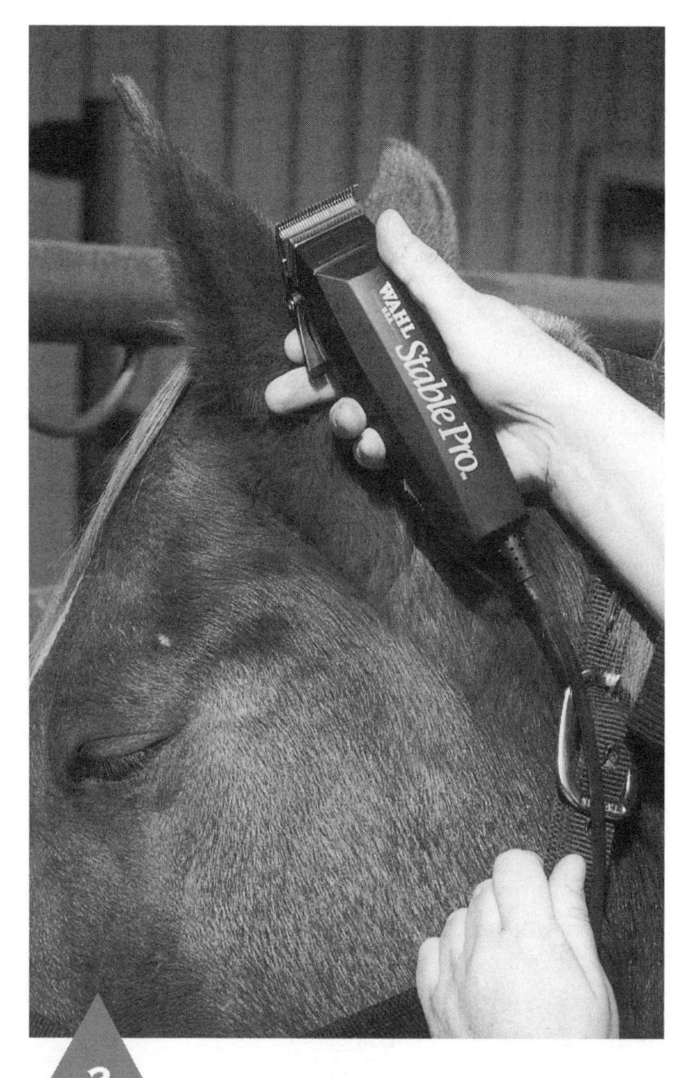

2 Hold the running clippers against the horse's shoulder with your hand between the clippers and the horse as a mediator of the vibration. Then place the body of the clippers on the horse.

Before you get ready to clip near the poll or head, accustom the horse to the vibration in those areas.

Clipping a Bridle Path

A bridle path is a lock of hair that is removed between the forelock and the mane. Clipping a bridle path provides a tidy place for the crown piece of a halter or bridle to lie without getting tangled in long mane hair.

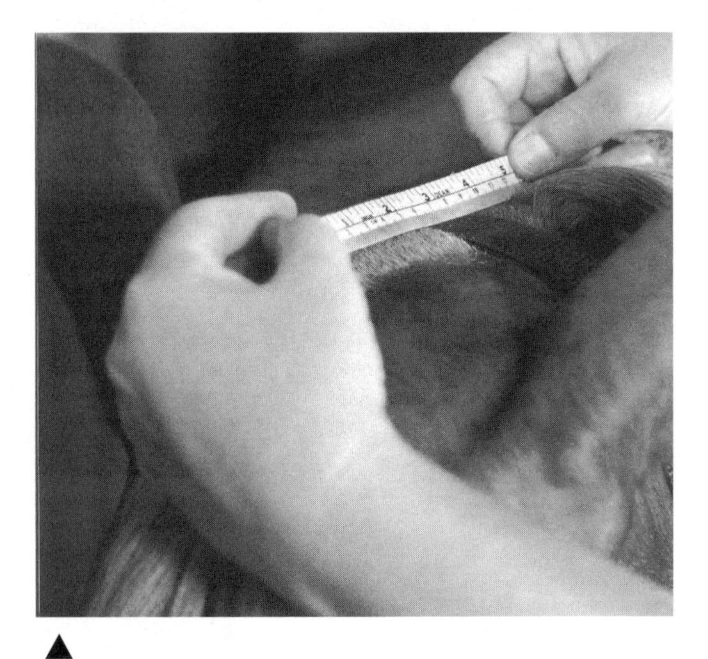

▲
Length of the Bridle Path
The length of the bridle path will be determined by the breed, use of the horse, and your preference. It's safest to make the bridle path as short as possible, as long as it is still functional. A 3- to 4-inch bridle path is about right for most horses.

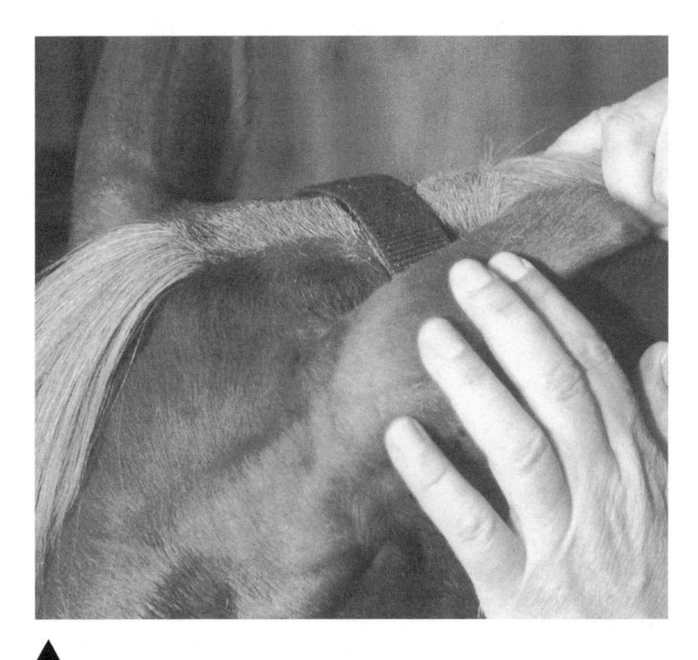

▲
A Rule of Thumb
One gauge is never to make the bridle path longer than the length of that horse's laid-back ear. This horse's 4-inch bridle path stops short of the ear tip by about 1 inch.

1

Before clipping, slip the halter back in position so it holds the mane out of the way. You may have to loosen the halter a few holes in order to do this.

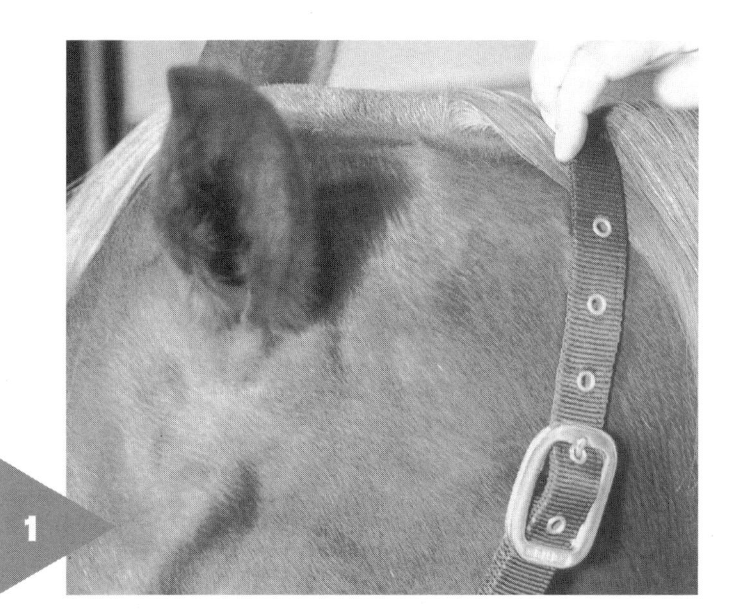

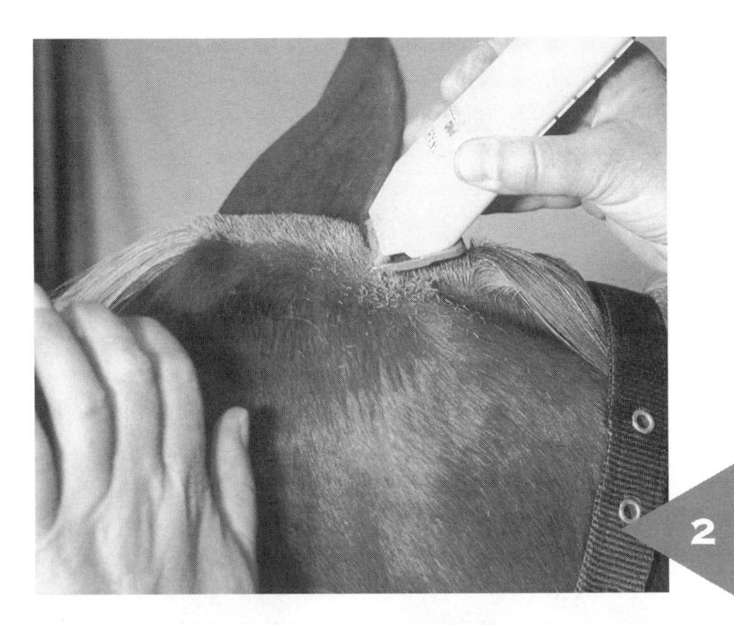

2 With your left hand, hold the near ear forward to keep it out of the way of the clippers. First clip from back to front.

3 Then clip from front to back.

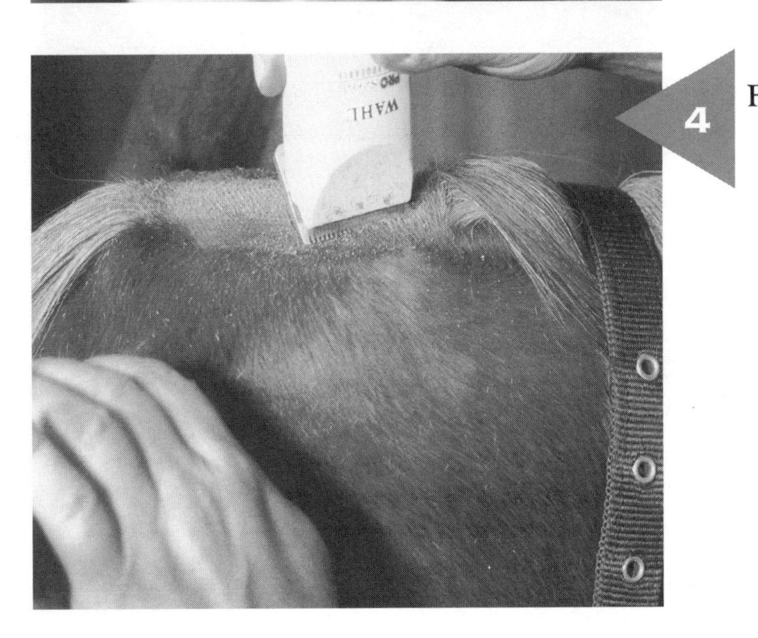

4 Finish by clipping down the sides of the bridle path.

Clipping the Legs

Before you begin clipping, review the signals you used during bathing that told the horse to keep his leg straight and not pick it up as you work.

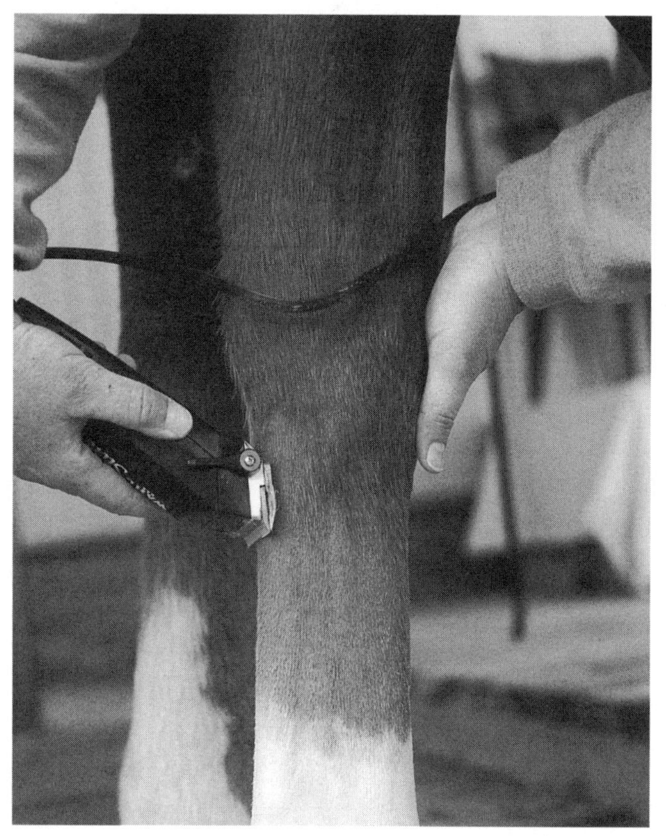

▲
Signals: Push Back on the Knee
When working on the front, place your hand on the front of the leg and push back on the knee.

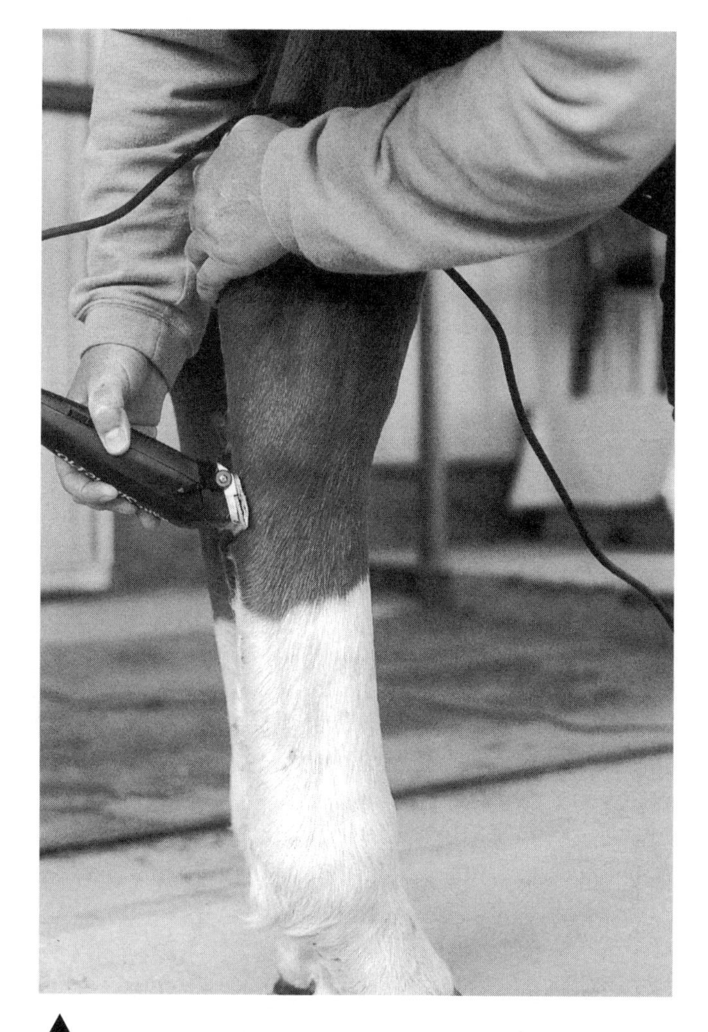

▲
Signals: Push Forward on the Hock
When you are working on the hind leg, push forward on the hock.

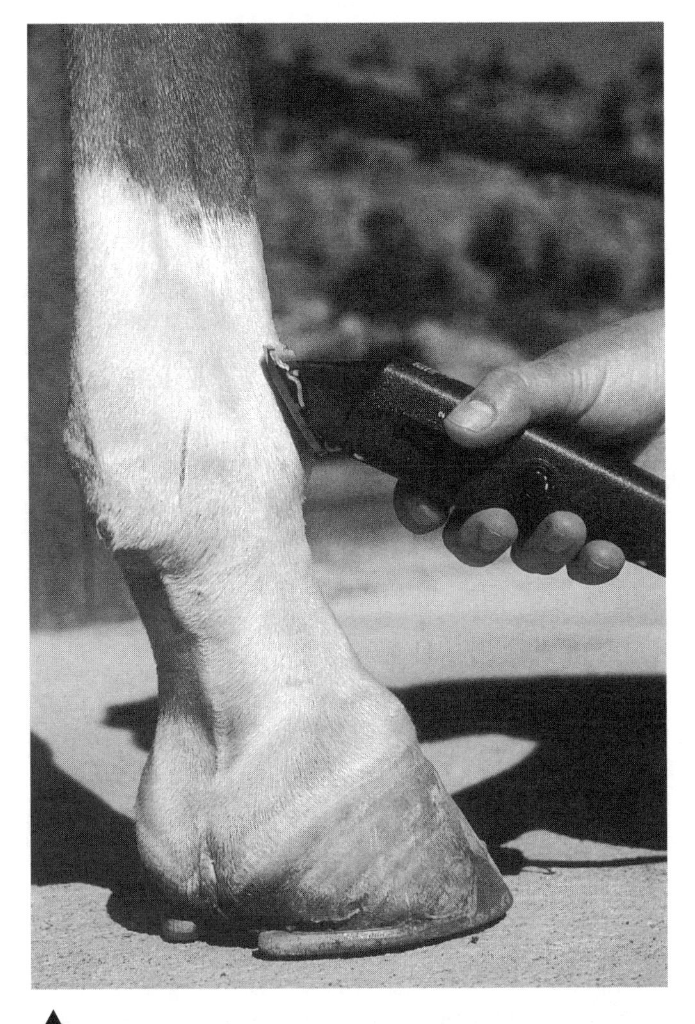

▲
SELECT THE BLADES FOR CLIPPING
For a cool summer cut or a show cut, you can use regular (#10) blades against the hair.

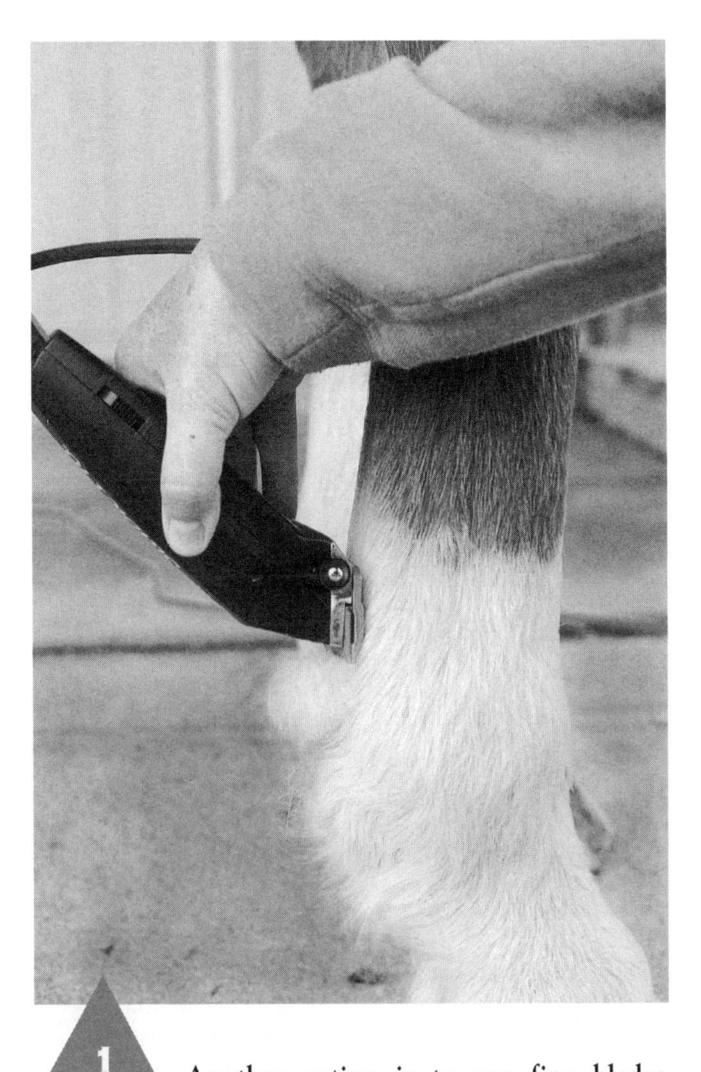

1 Another option is to use fine blades (#30) and clip in the direction of the hair growth. Start at the flexor tendon region. Keeping the blades flat against the leg and the pressure even, move the clippers slowly down the leg. Repeat, taking off a path of hair with each stroke.

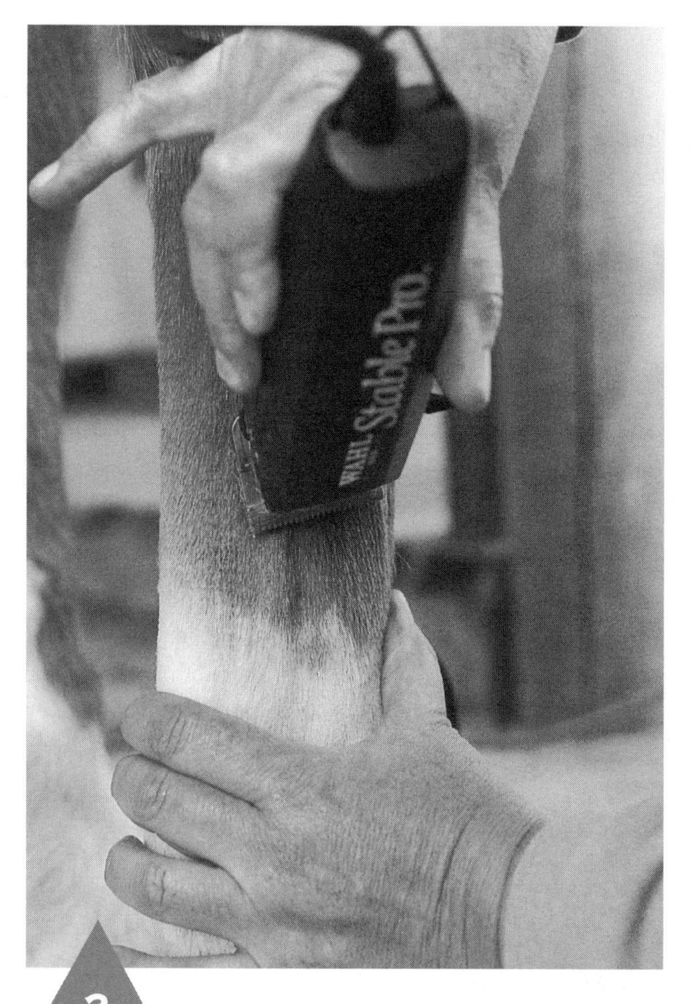

2 When you come to the outside of the leg, after you've made a few strokes you'll find that some tufts of hair hide in the channels formed by the flexor tendons, the splint bone, and the cannon bone. To get at this hair, roll the skin forward and it will bring the hair out of the channels so you can clip it.

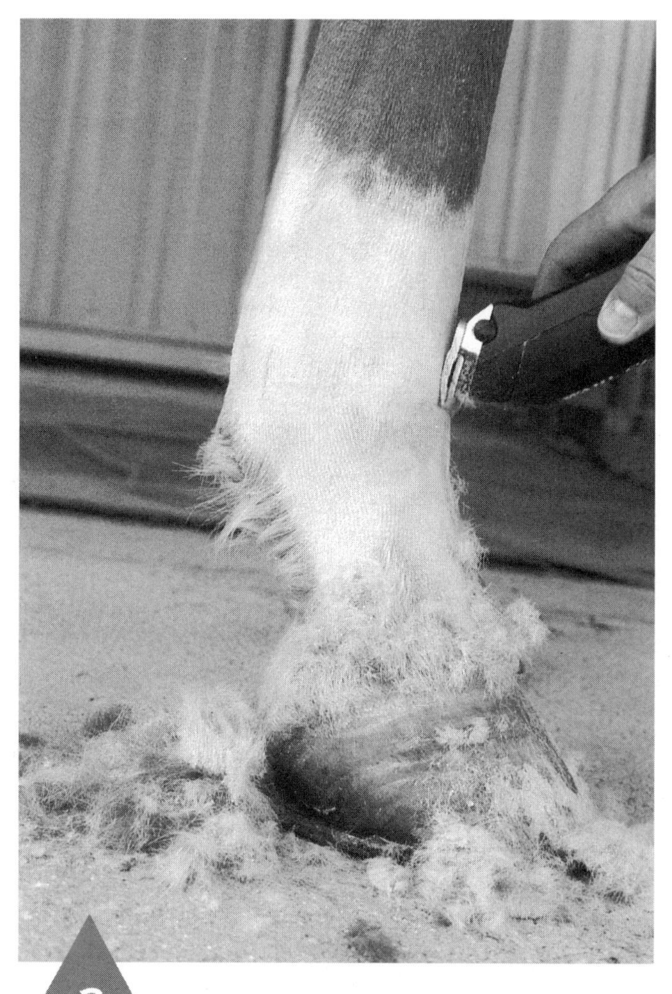

3 Continue clipping around the leg, keeping the blades at a constant flat angle to the leg.

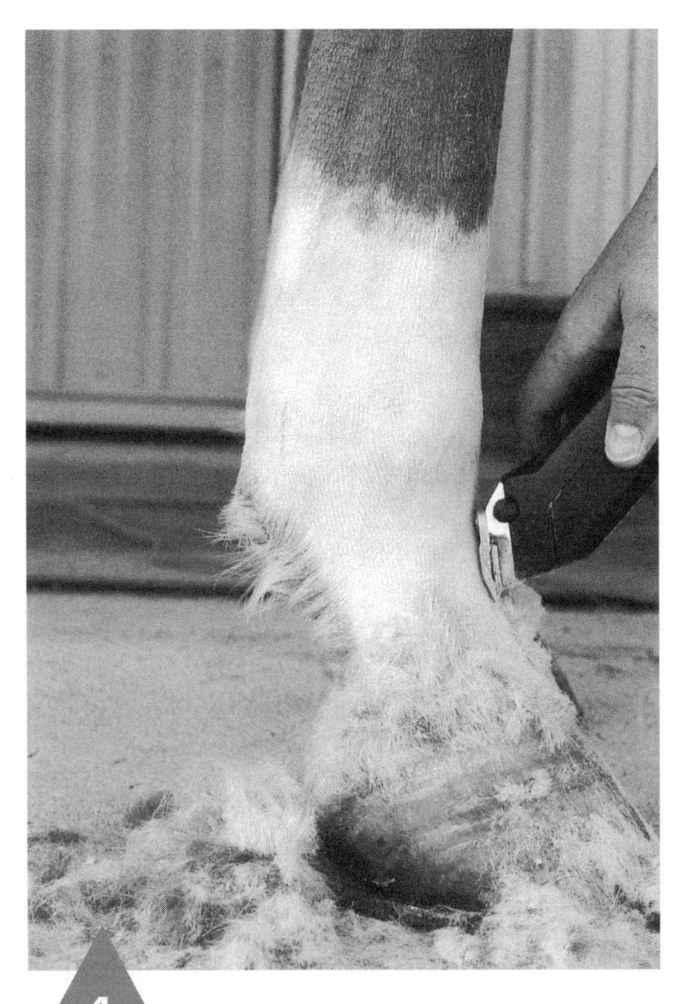

4 When you come to the curved portion at the front of the horse's pastern, be sure you follow the curve with your clippers or you will end up with ridges and furrows.

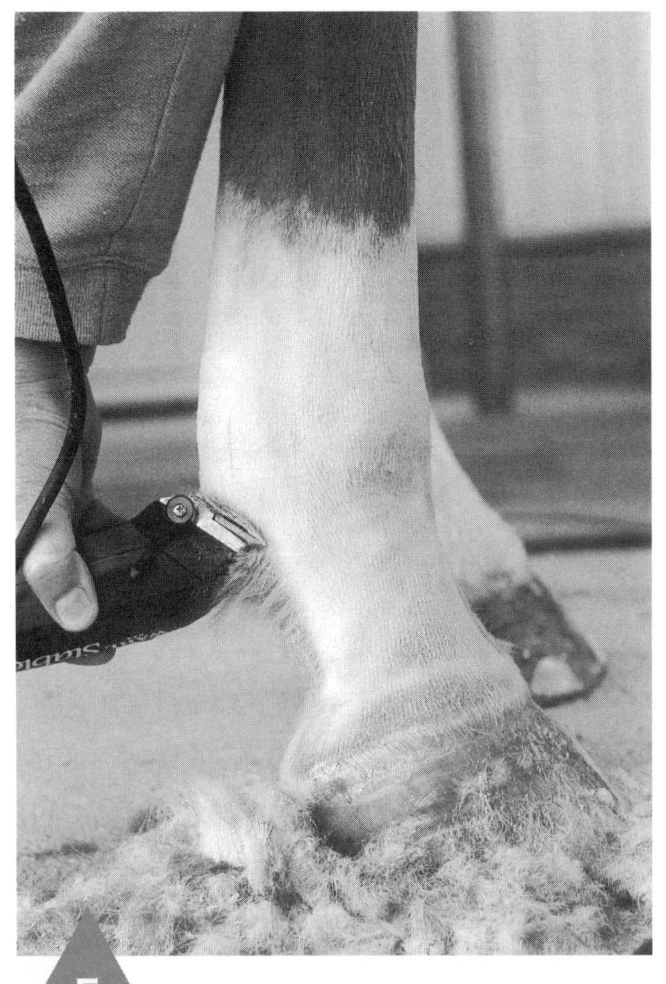

5 When clipping the back of the pastern, place the clippers under the ergot.

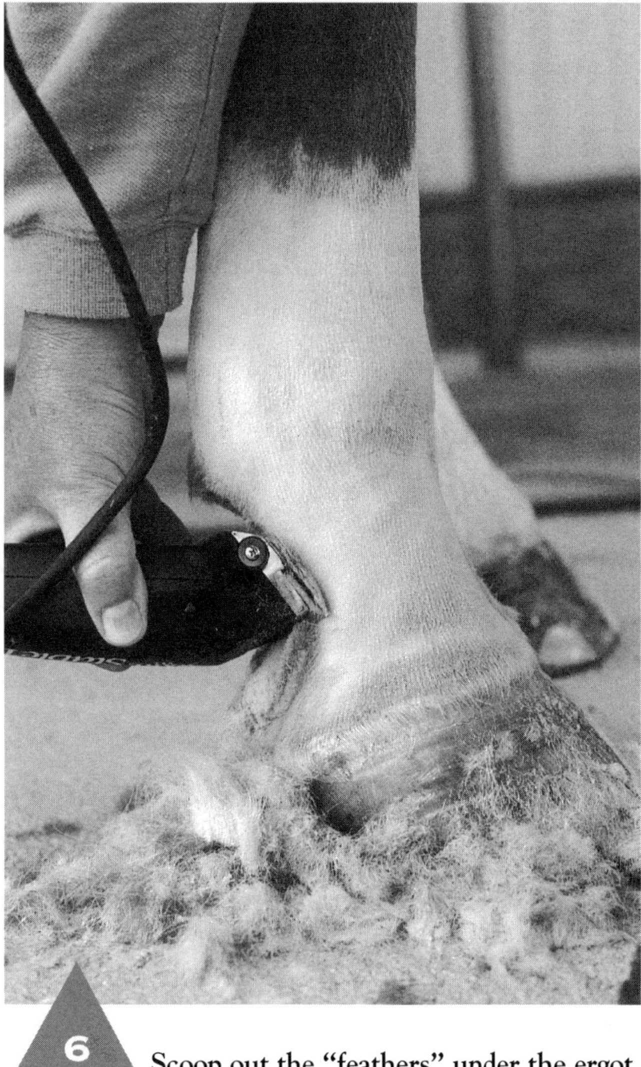

6 Scoop out the "feathers" under the ergot. A horse who is accustomed to clipping won't move a bit during the entire process. The result: a smoothly clipped leg and one pile of hair that is easy to clean up.

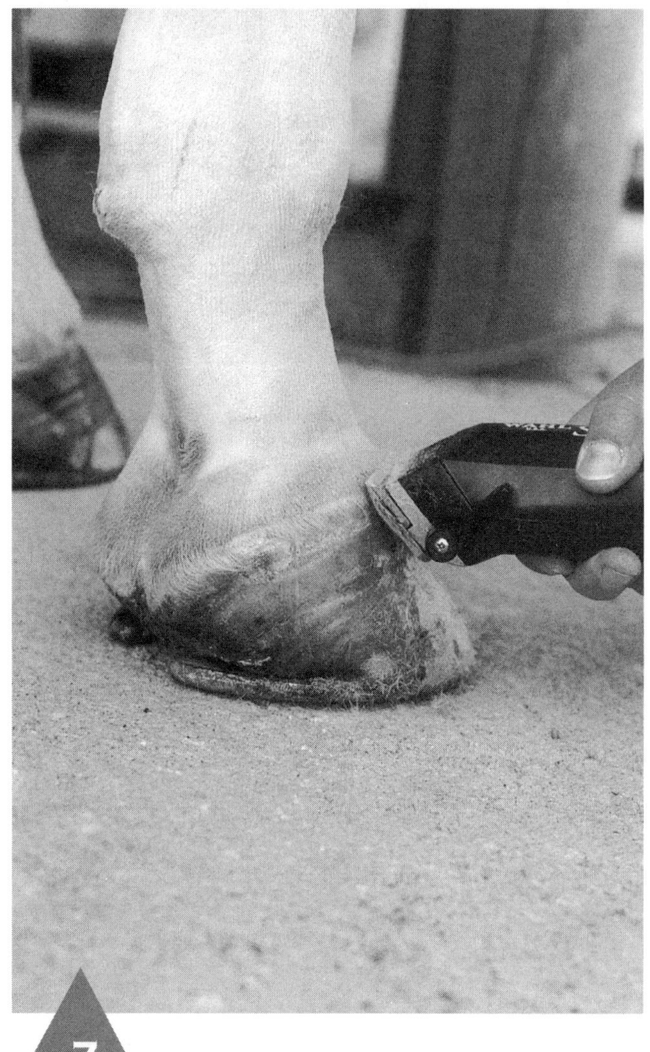

7 After you've cleaned away the pile of leg fluff, finish the clip job by clipping upward, against the hair growth, around the coronary band.

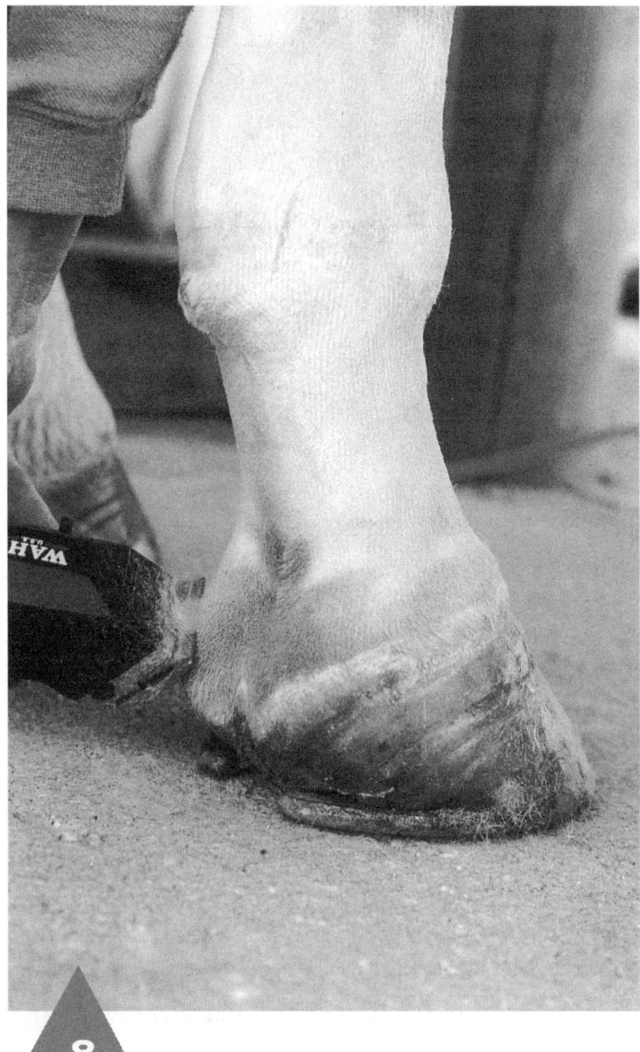

8 Don't forget to clip the area near the bulbs of the heels.

▲

Option: Hold the Leg

Sometimes it's easier to finish the trim job while holding the leg so you can extend the pastern joint, which opens it up for you to work.

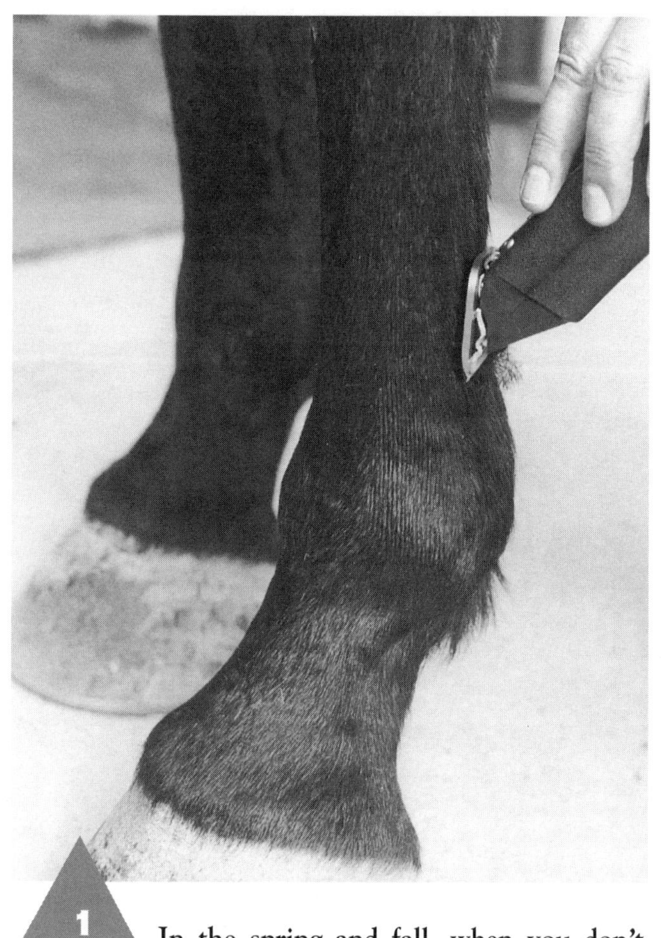

1 In the spring and fall, when you don't want to remove much leg hair but you still want your horse to look tidy, use regular (#10) blades in the direction of the hair growth. That way you won't "scalp" your horse's legs as you would when you use the same blades *against* the hair growth. As always, keep the blades flat and consistent for an even cut.

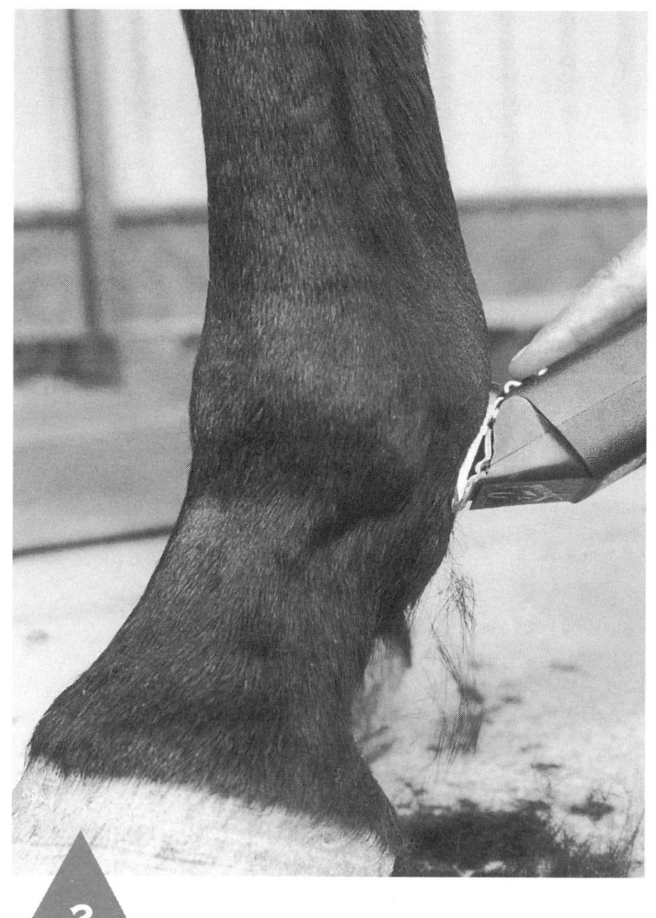

2 Scoop out the fetlock. Notice that very little hair is being removed. What you are basically doing is trimming the long hairs and evening out the hair coat.

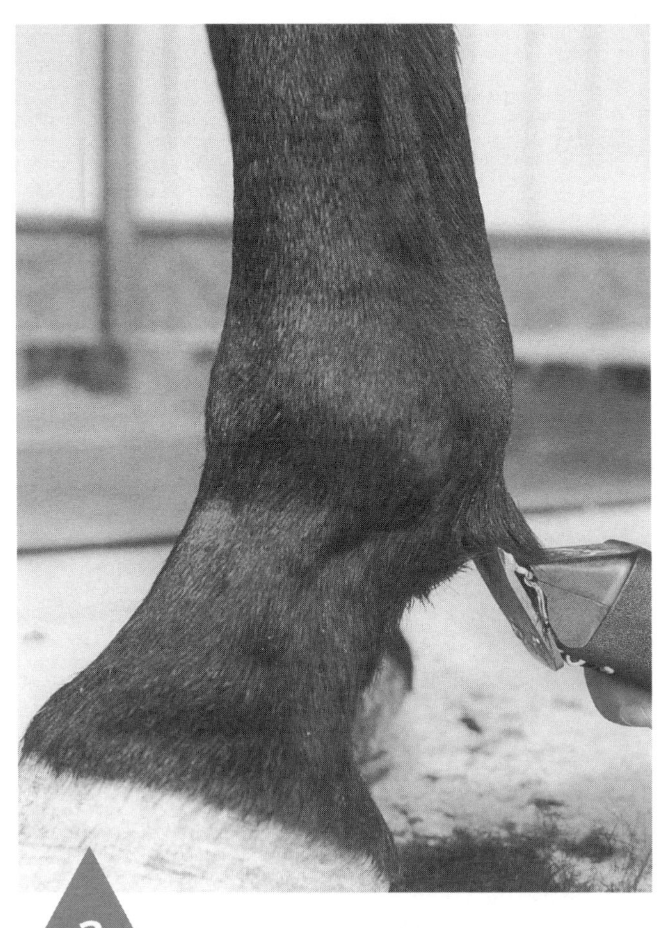

3 For the difficult-to-remove fetlock feathers, turn the clippers against the hair growth and, holding the blades away from the skin ¼ inch or so, clip upward.

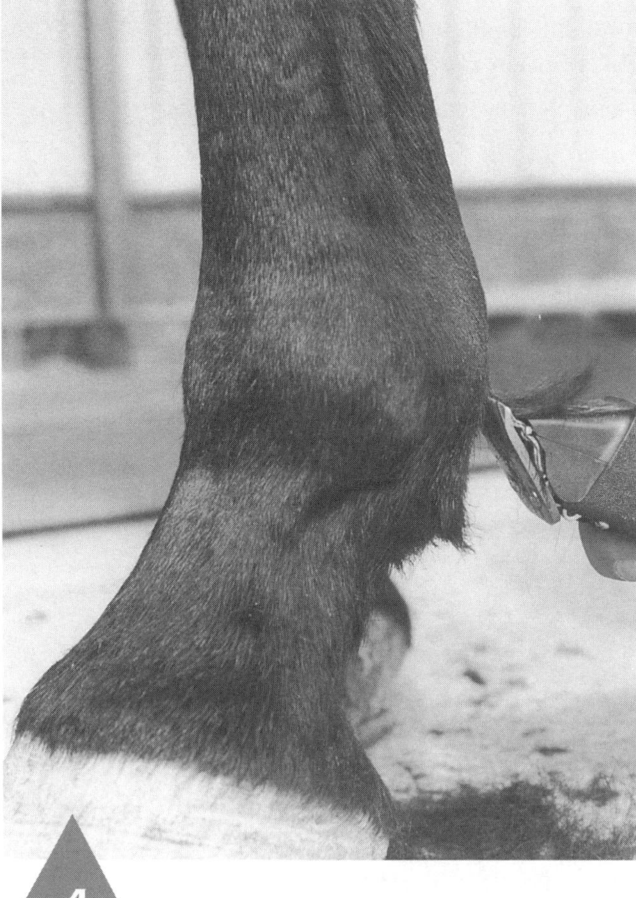

4 This will hinge the hair up and allow you to nip it off. Don't rest your clippers on the skin when you do this or your horse will end up with a bald patch.

Trimming the Ears

After fly season, if your horse won't be turned out for the winter (that is, you will keep him in the barn) and you want to tidy his appearance a bit, you can level his ear hairs.

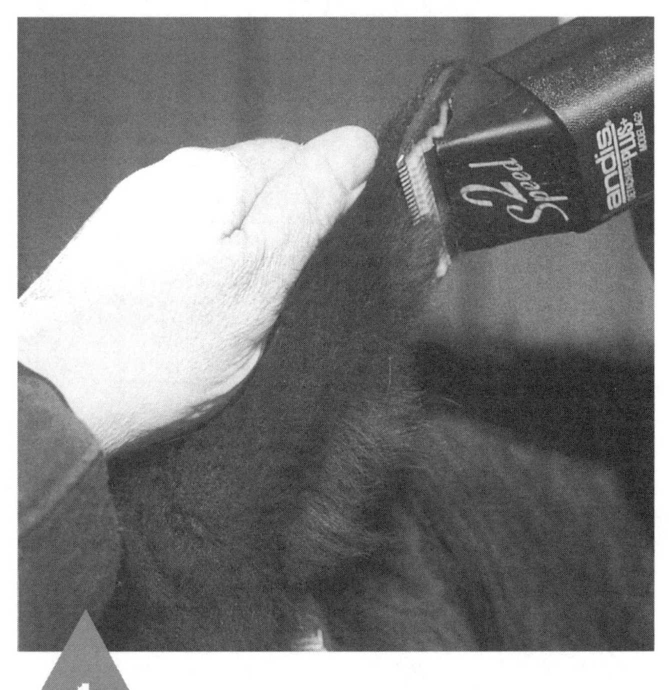

1 Fold the ear in half so the edges meet evenly. Run your clippers (with a #10 blade) along the edge and nip off any hairs that protrude from the ear.

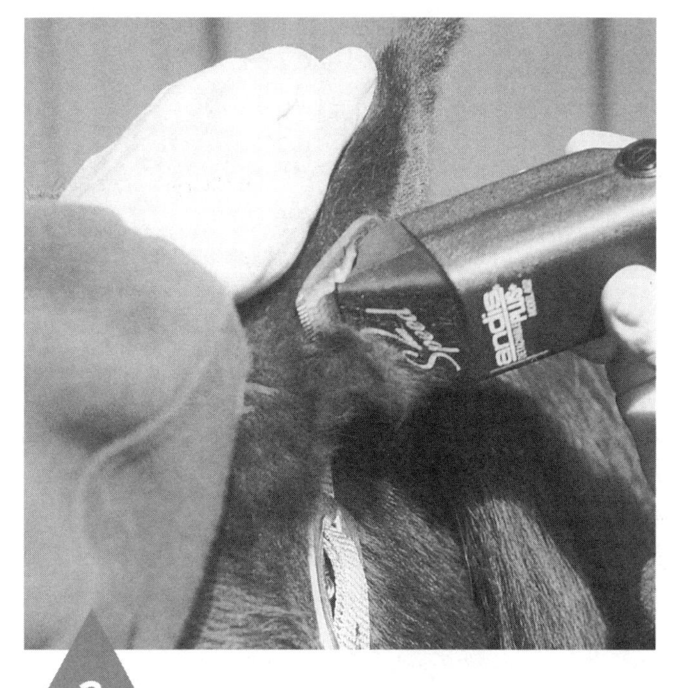

2 Continue clipping downward until the fluffy tufts have been removed.

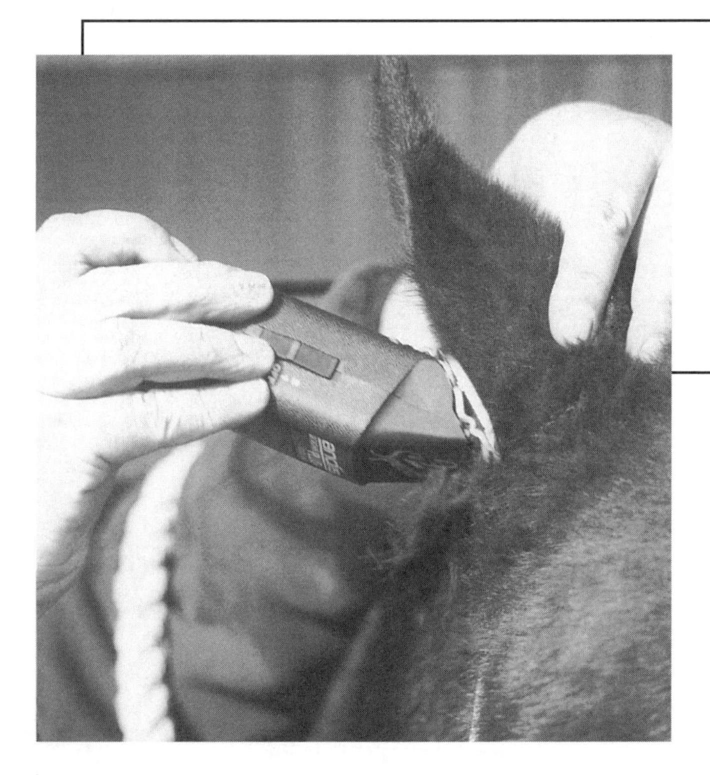

Trimming Tip

Always keep your blade flat. Besides resulting in an even cut, this will prevent you from accidentally nicking the edge of your horse's ear with the clippers. It doesn't take this happening more than once to make a horse shy of ear clipping.

3 Let the ear open about halfway and get some of the longer hairs that pop out. Put your clippers down and massage your horse's ear, flicking the tiny little hair clippings out of his ear and fluffing out any long stragglers that might still be hiding.

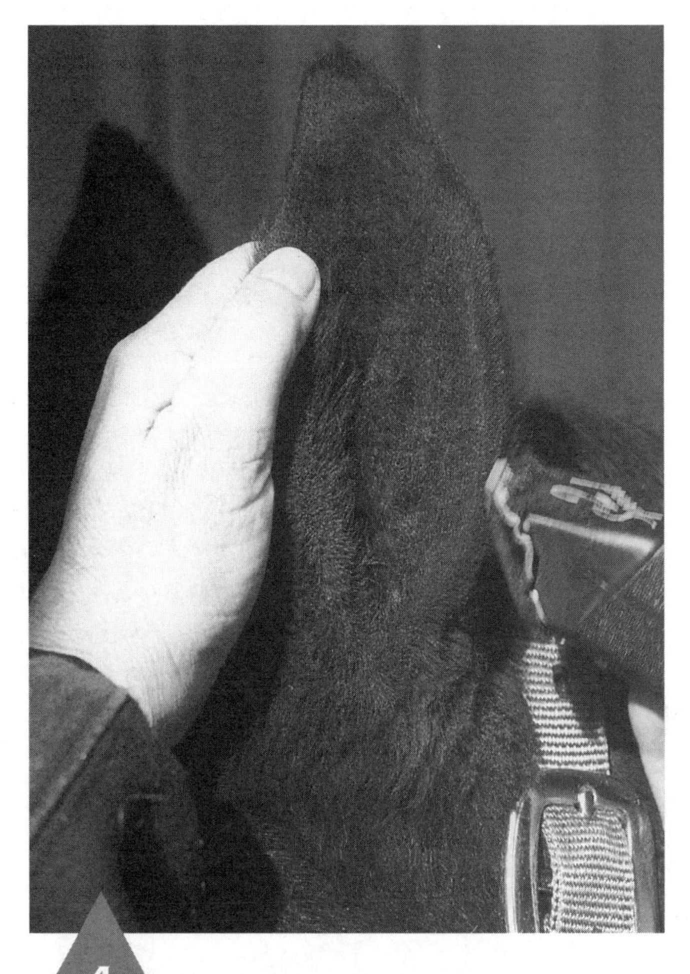

4 Run the clippers around the edges of the open ear to remove any long hairs that detract from the shape of your horse's ears. Now your horse's ears are tidy but still moderately protected.

Trimming the Jaw

Another place you might want to trim is under your horse's jaw. The long hairs here are a great protection against winter winds, so if your horse will be turned out, leave them intact. However, if your horse will be in regular work during the winter and you plan to keep him up at the barn, you can remove the guard hairs from his jawbone.

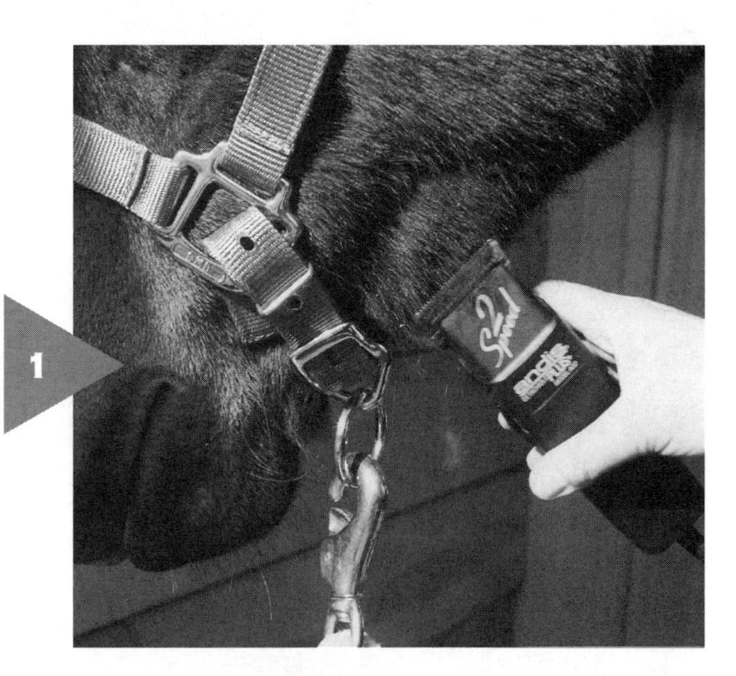

1 It is easiest to clip a horse's jawbone if he is wearing a grooming halter because there will be no throatlatch in your way. With the blades (#10) not touching the horse's skin but just hovering, make passes from under the jaw to the edge of the jawbone.

With the clippers still hovering over the horse's skin, make short passes from the edge of the jawbone up toward the horse's face, just removing the guard hairs, not letting the blades dig any deeper.

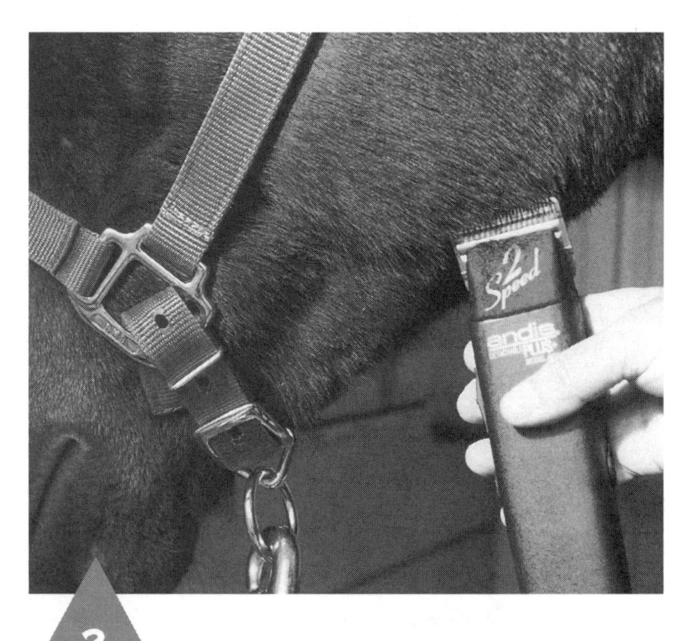

2 Make short passes along the cheek line with the clippers hovering over the horse's skin.

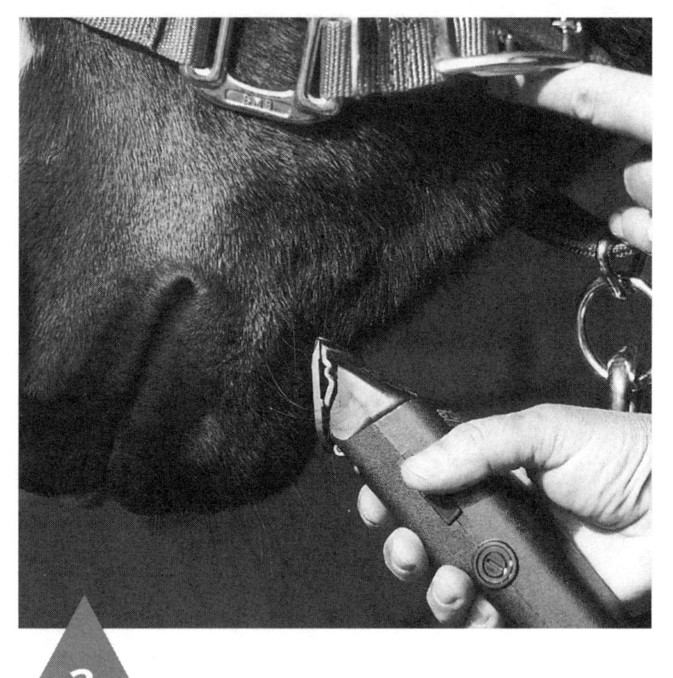

3 Loosen the noseband of the grooming halter and move it up out of the way so you can scoop out the chin groove, again with the blades hovering over the horse's skin, not touching.

Clipper Maintenance

Maintain your clippers between use so when you're working with your horse, you can focus on safe handling and creating a tidy appearance.

◀ To keep your clippers running smoothly and sharply, you will need various supplies for their maintenance. On the left are four types of lubricants: both clipper oils and grease.

- ✦ Lubricants should be used regularly but not heavily. Just a drop of oil on the blades is all that is necessary, but blades should be oiled before they are stored and while they are being used.
- ✦ The tall can is a spray coolant for use when you have been clipping for a long period of time and the blades are getting warm. After blades have been sprayed with coolant, they should be oiled. (*Note:* Not all clippers require coolant.)
- ✦ Blade wash is used to clean dirt, scurf, and hair from the blades. However, it is best to clip an absolutely clean horse — so your blades shouldn't really get dirty. Dirty blades get dull very, very quickly. After blades have been washed with blade wash, they should be oiled. Some clipper manufacturers suggest washing clipper blades in oil.
- ✦ The final bottle represents clipper disinfectant, necessary if you share clippers or suspect a skin problem. Sometimes that is precisely why you are clipping, so you can get at the skin to treat a skin problem. That is a situation in which a blade disinfectant is necessary.

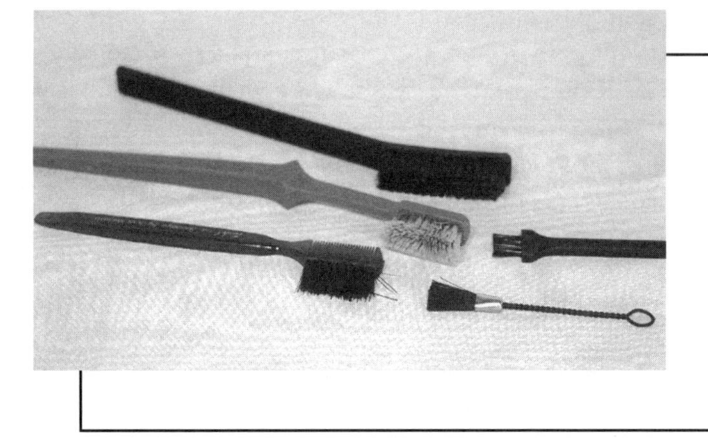

AS YOU CLIP and when you take your clippers apart to service them, you will want a variety of little brushes so you can get at all the nooks and crannies that tend to harbor dirt and hair. Besides the brushes that come with new clippers, you can also add some old toothbrushes or eyebrow brushes to your clipper maintenance kit.

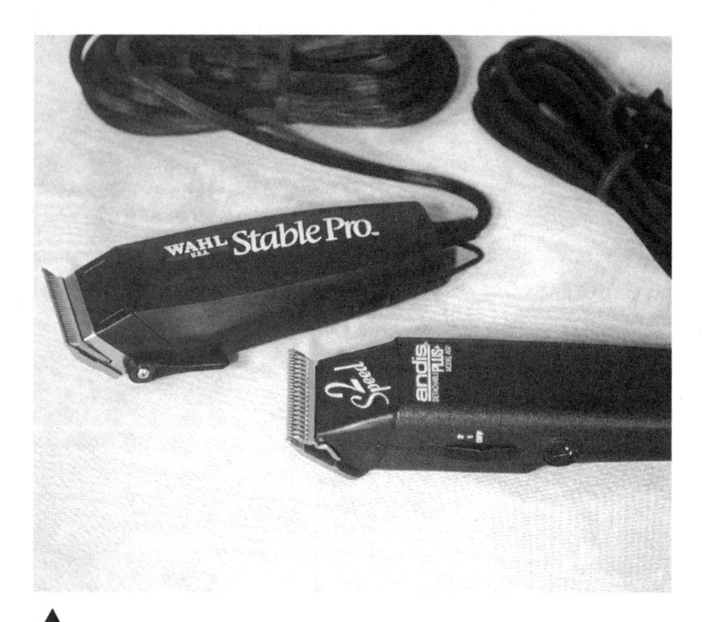

▲
Two Types of Clippers
The single-speed Wahl Stable Pro clippers have screw-on blades that can be adjusted to clip from size #10 to size #30. The two-speed Andis Plus has a #10 flip-off blade set that can be exchanged for a blade set of another size.

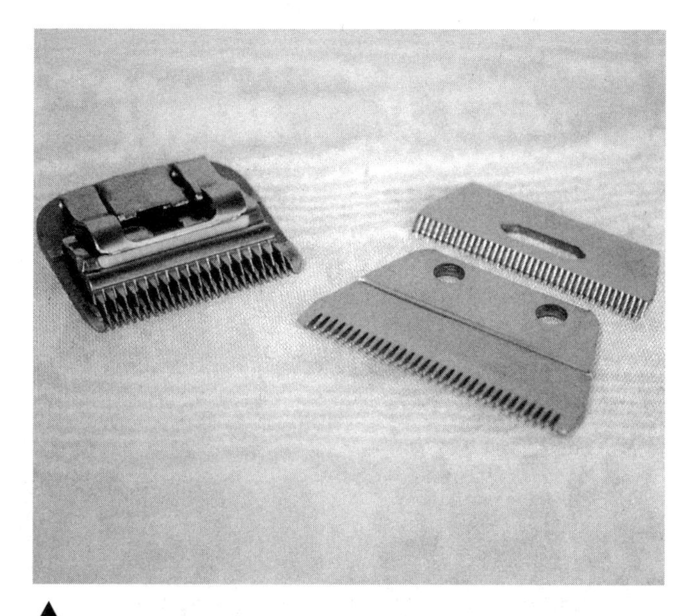

▲
Blade Sets
The Andis flip-off (left) and the Wahl screw-on (right) are two types of blade sets.

◀ **Rechargeable Clippers**
Besides traditional clippers, you may find rechargeable clippers handy. Because they don't require electricity to operate, you can take them to a horse at any site. This is especially convenient if you need to clip a wound and there is no electricity nearby or if you simply want to put a pair in your pocket to trim the bridle path of a horse in pasture or at a show. In order for rechargeable clippers to work well when you need them, clean and oil them as you would any other clipper and follow the charging instructions that come with the clippers.

▲
CLEANING CLIPPER BLADES
Clean dirty clipper blades in blade wash or oil. With the clippers running, dip only the blades in the fluid. It is handy to have a shallow container of oil nearby as you clip. Keep the container covered when you are not using it.

▲
WORKING WITH WARM BLADES
When you have been clipping for a long time, certain blades might get warm. If so, after you've cleaned them in blade wash, spray them with a cooling spray. Then be sure to oil them.

Removing Flip-Off Blades

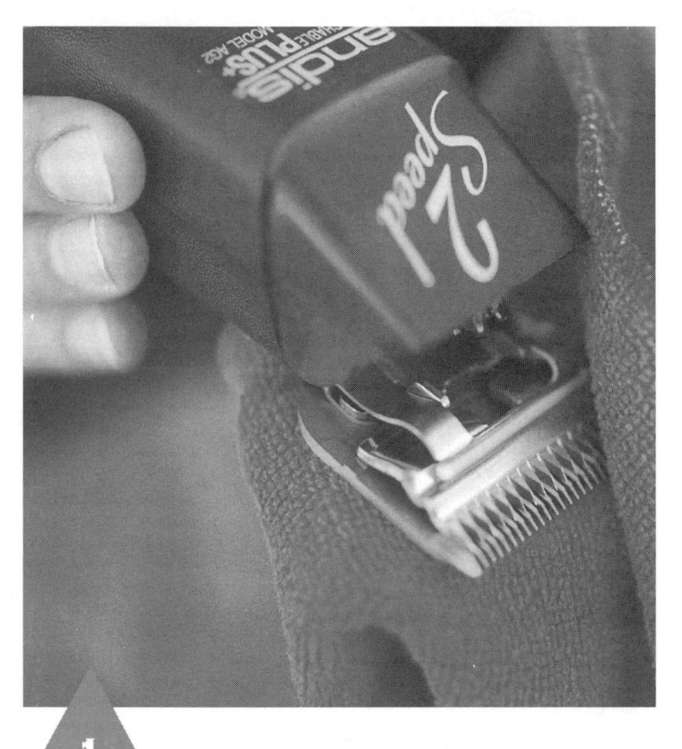

1 With some clippers, to remove the blades, just flip the blade set backward and slip it off. The top and bottom blades come off as a unit.

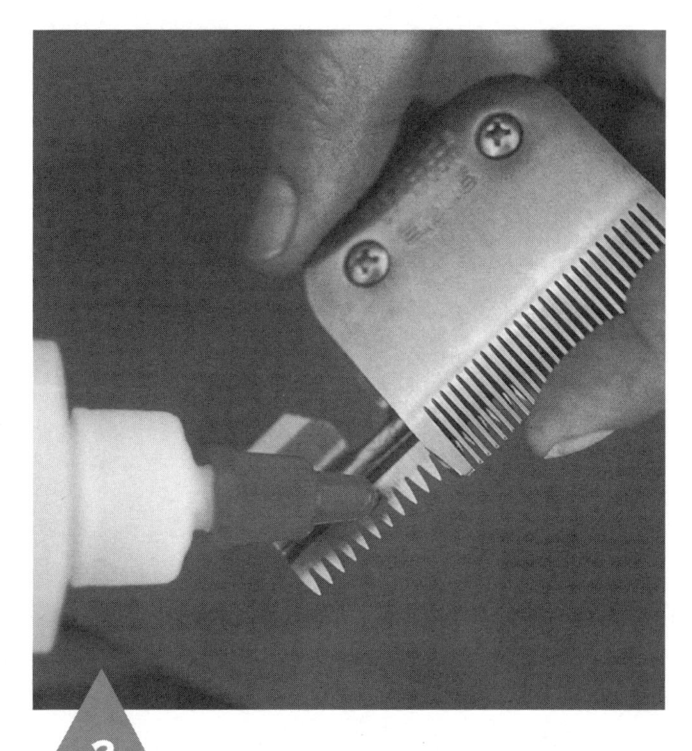

2 Slide the upper blade off to one side so you can place a few drops of oil on the runners that come in contact with the lower blade.

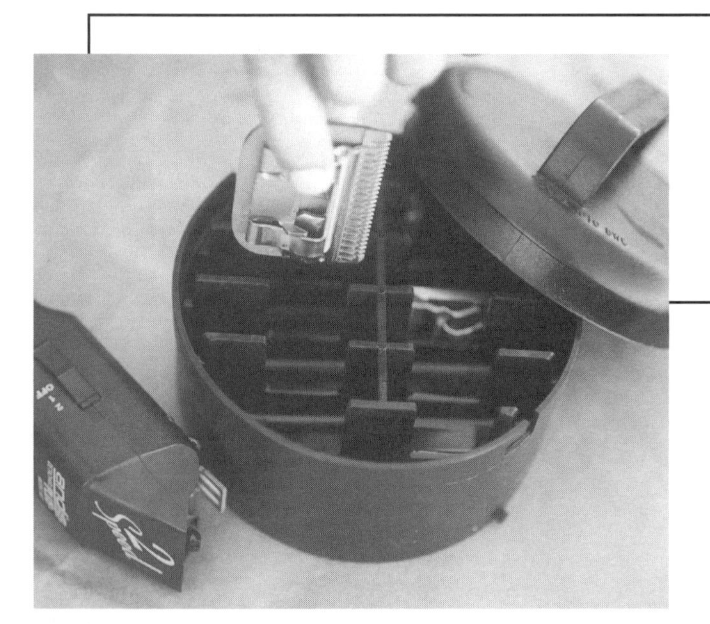

Blade Storage

You can store clipper blades in a blade caddie, which is a covered plastic container with individual cells for various blades. The container is filled with blade wash or oil.

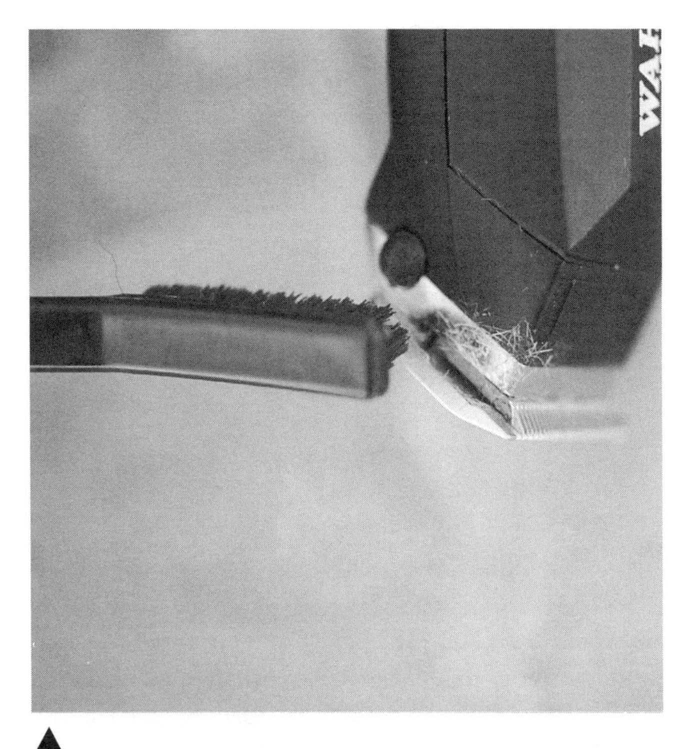

▲
QUICK-CLEANING THE CLIPPING HEAD
As you clip, use a brush to clean hair off the clipping head. You can use the brushes that come with the clippers, or a toothbrush will work just as well.

▲
REMOVING RESIDUE
If a residue builds up in the cutting teeth, you can help dislodge it with a small brush. Ultimately, however, you will have to clean the blades with blade wash or oil.

Removing Screw-On Blades

1 When you need to remove screw-on blades, rest the clippers on a table so that the blades don't suddenly drop. This could break the cutting teeth, making the blades worthless.

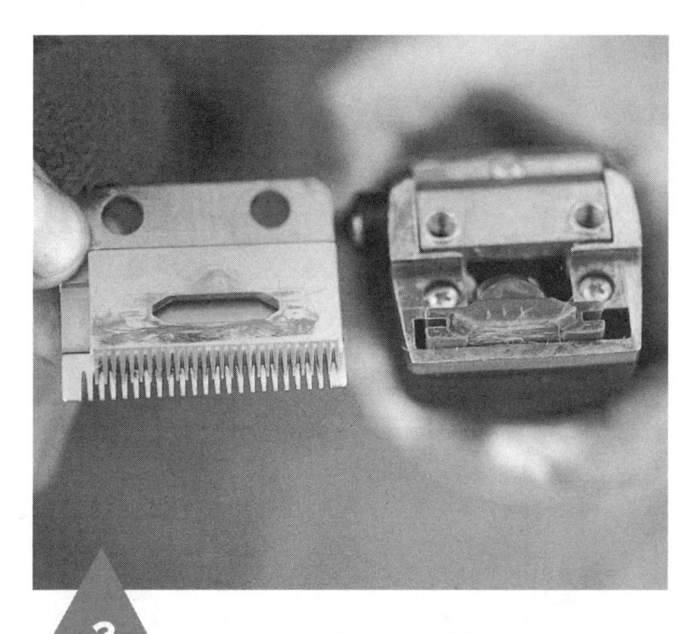

2 Remove the blades and clean out any hair buildup under the blades. Use a brush to reach all the nooks and crannies. Remove the blades and clean out any hair buildup under the blades. Use a brush to reach all the nooks and crannies.

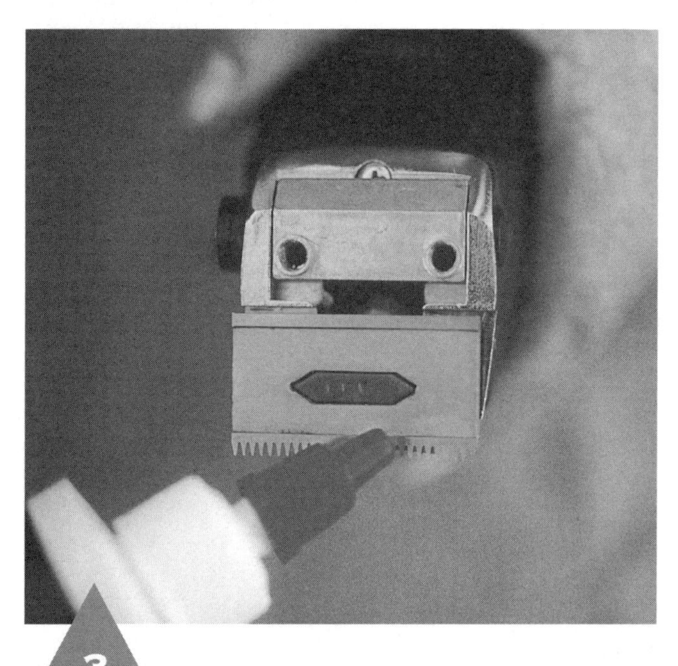

3 Oil the lower blade before you put the clippers back together.

The Care of Large Clippers

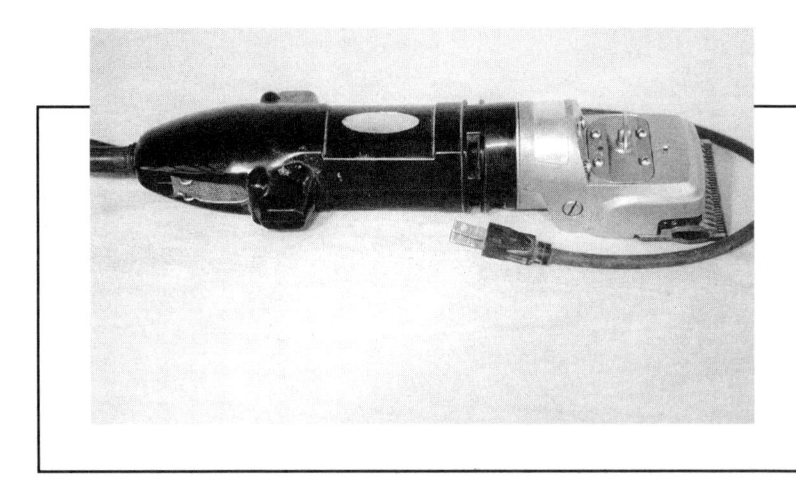

For Difficult Jobs

If you do a lot of heavy-duty winter coat clipping, you might use a large pair of clippers, such as these. These are very heavy and fatiguing to use for long periods of time but are tough workhorses for difficult jobs. They require the same care as the clippers previously described, with some additions.

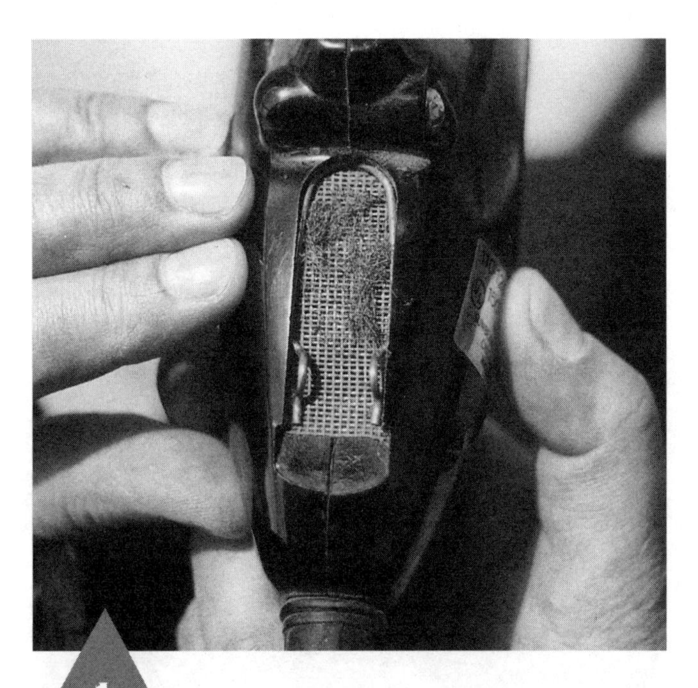

1 Keep the air intake screen clear of hair. When you are in the middle of a long clipping job (such as a body clip, in which all of the horse's hair is removed), you might not notice it, but hair piles up on the air intake screen, which essentially cuts off the air supply to the motor. When this occurs, the clippers can get very hot quickly.

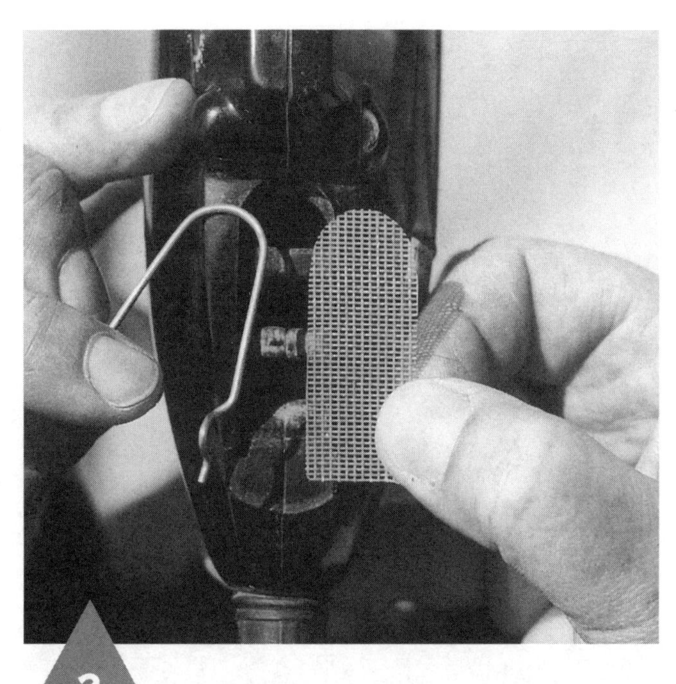

2 Remove the spring clip and screen and clean thoroughly before putting back together.

3 Remove the blades so you can do a thorough job of cleaning under them.

4 The area under the blades is a trap for hair. The hair becomes packed tightly in the crevices and causes overheated blades and inefficient clipping.

5 After digging out all you can with a screwdriver, blow out the rest with compressed air.

6 Oil the blades before you put them back on. Also add a drop of oil to each oil hole on the top of the clippers.

7 Adjust blade tension with the set screw. Clockwise tightens; a counterclockwise turn loosens tension.

Caring for the Mane

UNTANGLING A LONG MANE

1

If your long-maned horse is turned out on pasture, especially during a windy season, it won't take long before the mane is tangled horribly. Add a few burrs or thistles, and you might be tempted to shave the whole thing off. But don't give up. This mane can be saved.

Coat the entire mane, especially the sections tangled the most, with detangling spray or gel. Work it in with your hands (without adding more tangles.)

2

Wear a pair of old leather gloves. The detangling product will build up and make a glassy coating on the gloves that will help your fingers slip through the mane more easily. Find a section that has become entwined. Usually what happens is that two sections twist and roll inward and form a circle. After examination, you will likely discover a pattern and find out which way you need to untwist it to untangle it.

Once you have untwisted the major tangles, start fingering through the mane with your gloved hands. You can add some more detangler at this stage. Try not to pull when you come to tangled spots — this will just end up in hair loss and be uncomfortable for the horse. Although it doesn't seem to hurt a horse very much when we pull out a few mane hairs, when you tug hard on a tangled mane, it pulls on the crest of the neck, which most horses quickly learn to dislike. So take your time and work out any twists, knots, and tangles.

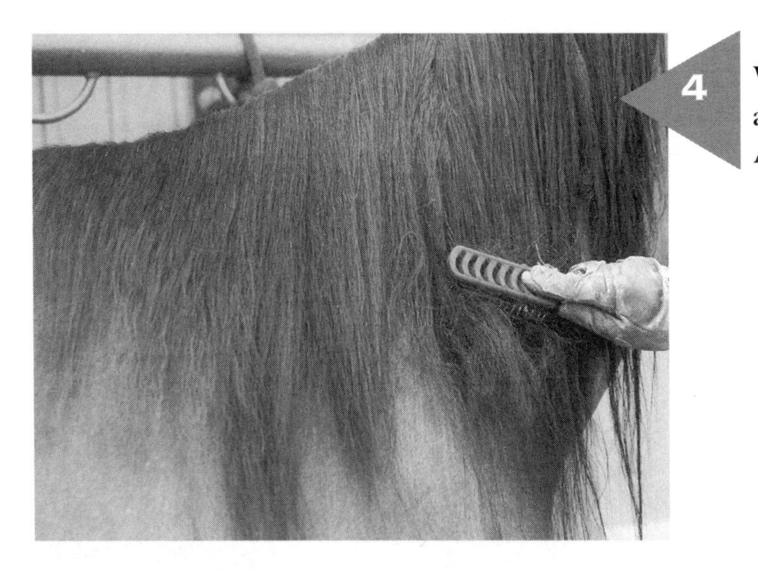

4

When it is time to begin work with the brush, start at the bottom of the mane and work your way up. As soon as the brush meets resistance, stop.

5

It might be a burr or other plant material that needs to be removed.

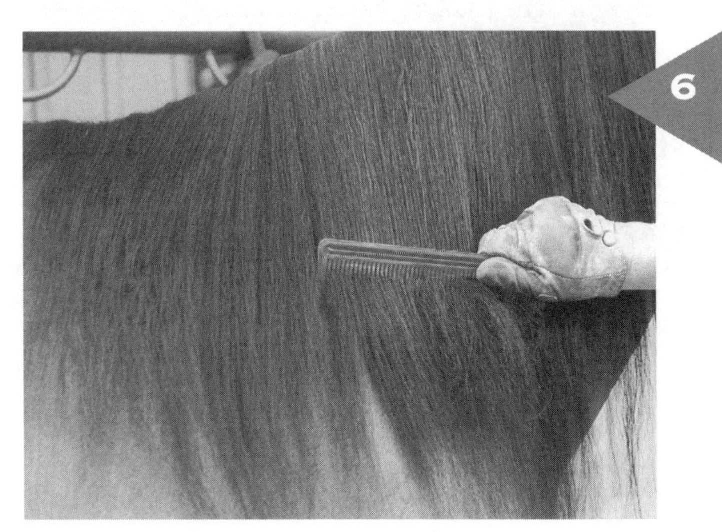

6

Only after you have successfully passed through the mane with the brush several times should you attempt to comb it. Now that the mane is untangled, treat it regularly with a detangling product and the hairs will be less likely to hold knots and burrs in the future.

Shortening the Mane

If your horse's mane has grown so long that it is out of fashion for his sport or if you have decided to change your long-maned horse to a short-maned horse, here's one way you can go about cutting it.

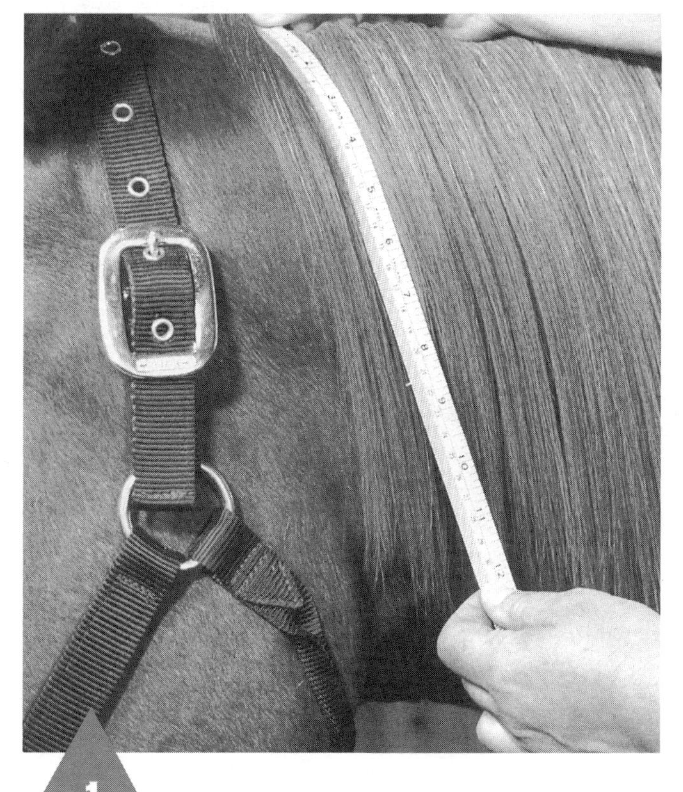

1 Start with a clean, well-brushed mane. To decide how long you want the mane to end up, hold a tape measure at the crest and estimate what a 4-inch mane, for example (a common length for both English and Western horses), would look like on your horse.

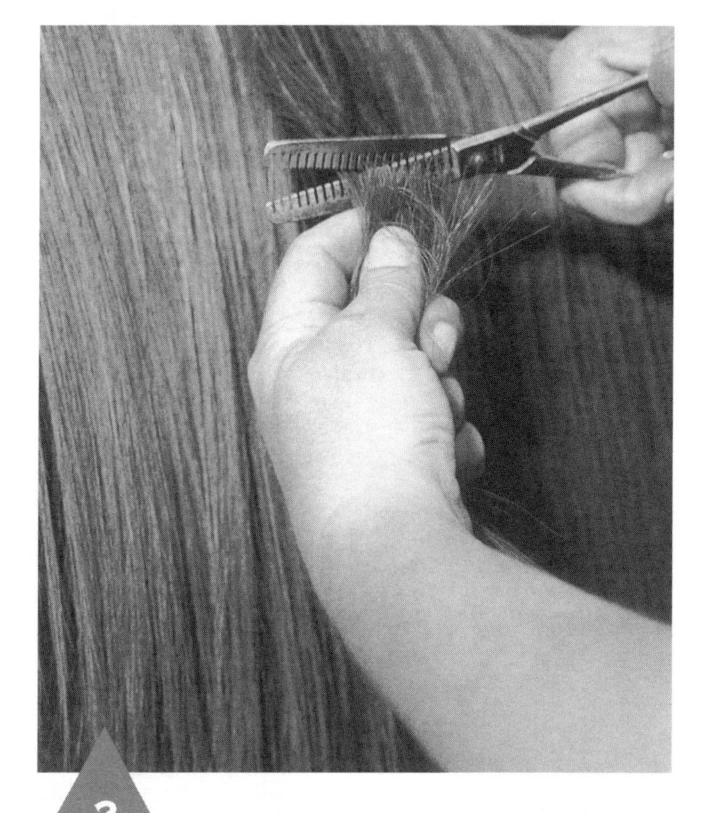

2 To make the first cut, which will remove the majority of the long mane, start in the middle of the mane, never at one end or the other. Take a section of mane and cut through it with a pair of thinning shears at a length about ½-inch longer than what you want to end up with; this is just a safety margin. The thinning shears will result in an irregular, more natural-looking edge than if you had used regular scissors. The irregular edge is desirable at this stage. You do not want a blunt cut.

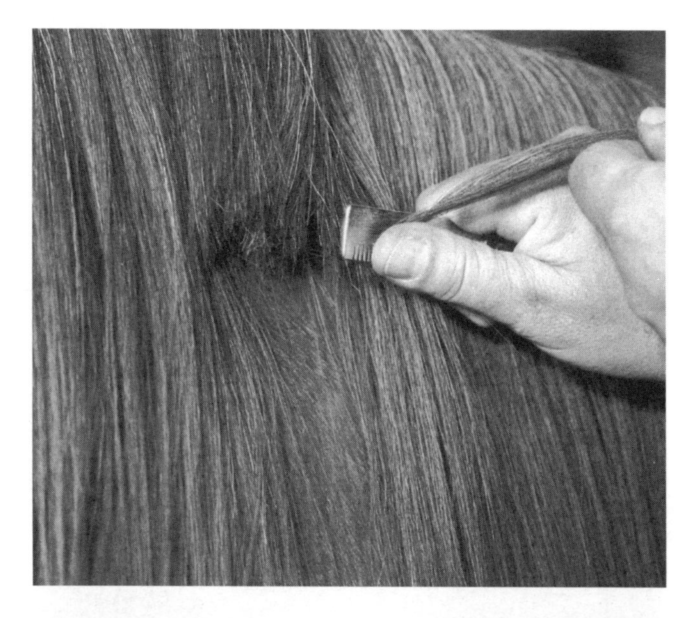

◀ Use a Thinning Comb

Another way to do the initial shortening is with a thinning comb, which is a small comb with closely placed teeth and sharp cutting edges between them. Take a small section of hair, place it between the comb's teeth, and pull down sharply on the comb. The hair will break off in a desirable, irregular edge.

◀ Use Electric Clippers

An even faster method for the initial shortening of a long mane is with a pair of electric clippers. Carefully cut through the mane in sections, constantly checking the length you are aiming for. Note that whatever method you choose for the initial shortening, you will end up with an irregular edge at this stage — which is what you want.

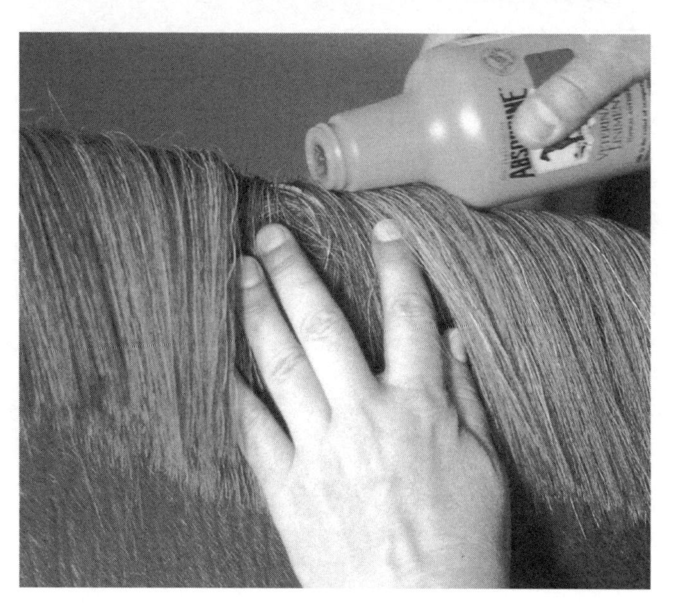

◀ Use Veterinary Liniment

Comb through the shortened mane and evaluate its overall length. It should be fairly consistent for the entire length of the horse's neck. Now apply some veterinary liniment to the crest of the horse's neck where the roots of the hairs have their origin. The liniment seems to lessen the horse's reaction to the mane pulling, which is coming up. It is best to put liniment on only about a 4-inch section at a time. Apply it fresh to each section as you work your way up and down the neck.

Thinning and Evening Up a Mane

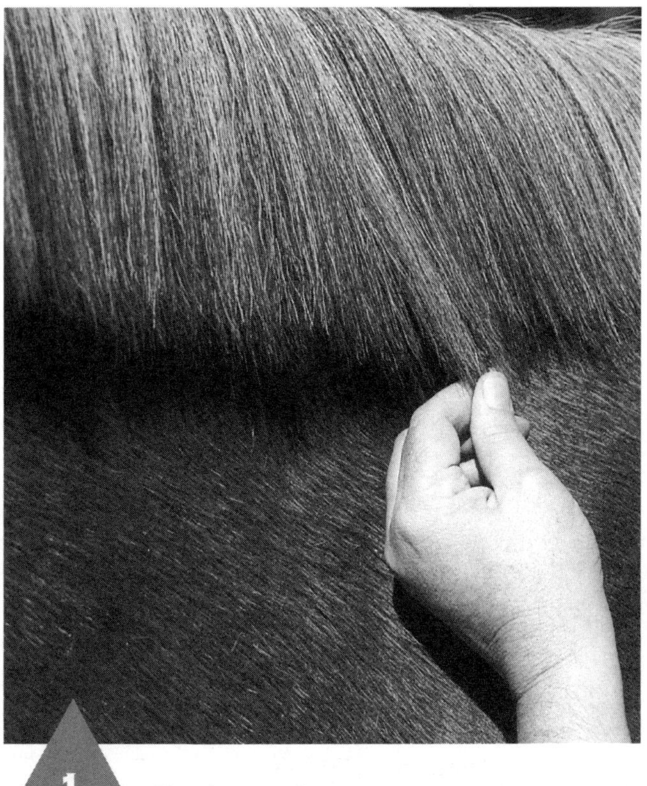

1 To thin and even up a mane, you can "pull" it. Select about eight hairs that hang down longer than the rest and hold the end of that group in your left hand.

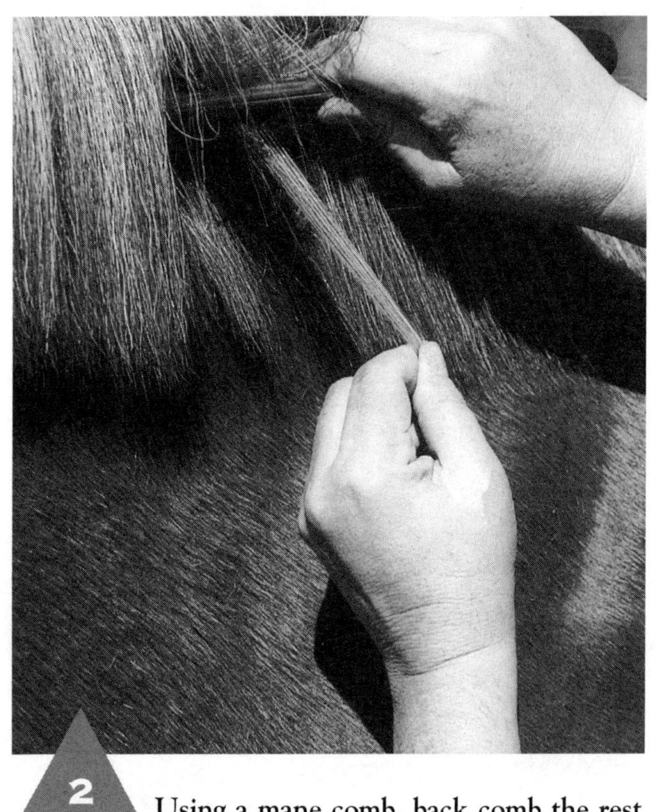

2 Using a mane comb, back comb the rest of the hairs away from the selected group, keeping the eight hairs firmly grasped in your left hand.

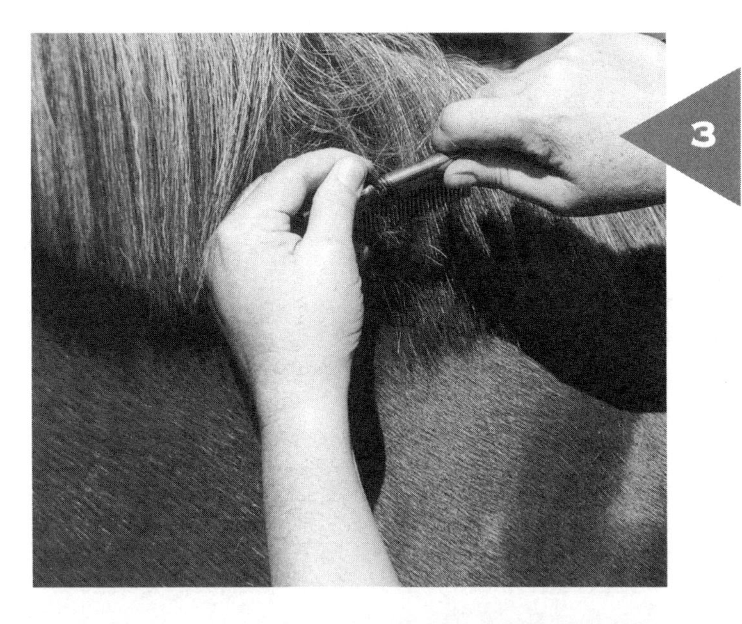

Wrap the group of selected hairs around the mane comb at least twice.

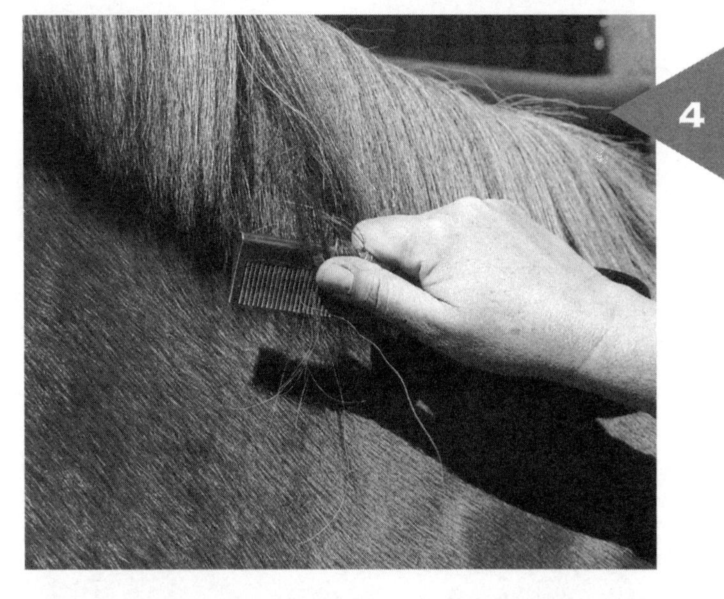

Hold the end of the wrap with the thumb of your comb hand and with one sudden pull remove the hairs. Most horses do not object to this process. If yours does, be sure you are not taking too large a group of hairs at once and be sure to use liniment. Also, your horse may be objecting to the pressure of his crest being pulled sideways. You can remedy this by holding your other hand at the crest to hold it upright as you pull.

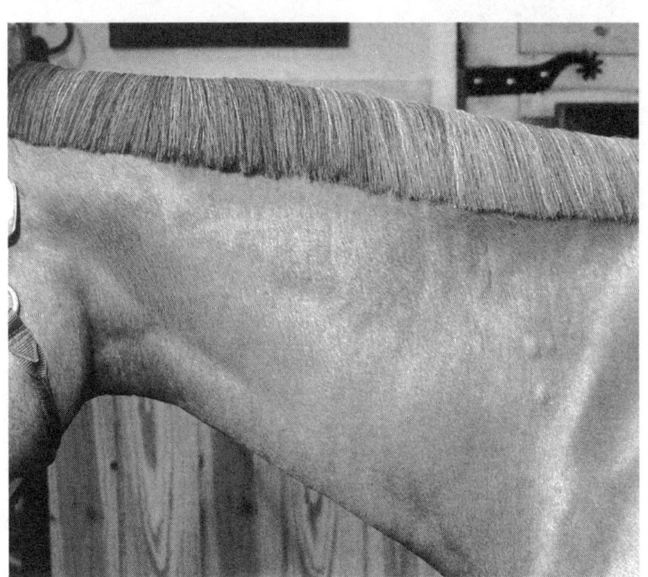

◀ **The Shortened Mane**
Here is the freshly shortened mane. In a few days, the hairs will relax and find their new place.

TRAINING A MANE BY BANDING

Many manes naturally fall to one side or the other of the horse's neck and stay there. Some manes, however, persist in frequenting both sides of the neck. To train the mane to lay flat on one side of the neck, you can use various techniques.

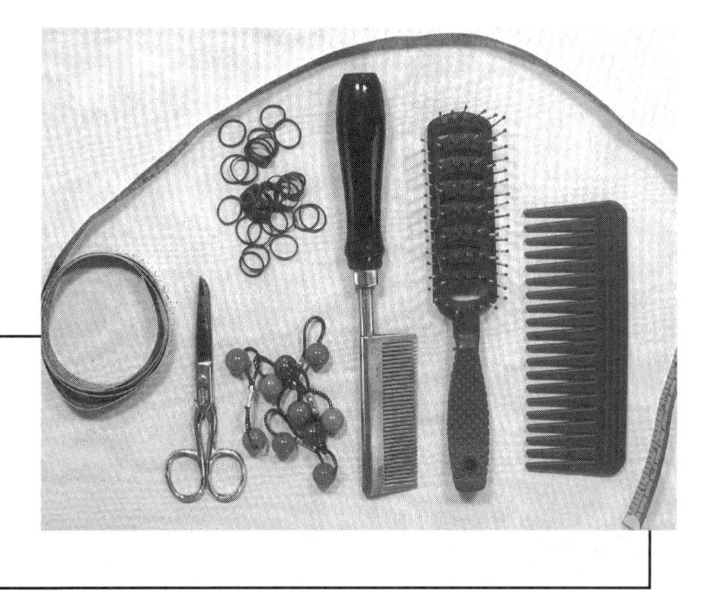

TOOLS HELPFUL IN mane and tail work include a measuring tape, small scissors, small rubber bands, double-ball ponytail elastics, a mane-pulling comb, a brush, and a mane comb.

1

Banding a mane is simply sectioning off the entire mane into 30 to 50 tiny ponytails that are fastened by small rubber bands, such as the ones used for show braiding. To band a long mane, take a small section of hair and separate it from the rest of the mane all the way down to its end.

2

Take that small section of hair and place a small rubber band around it.

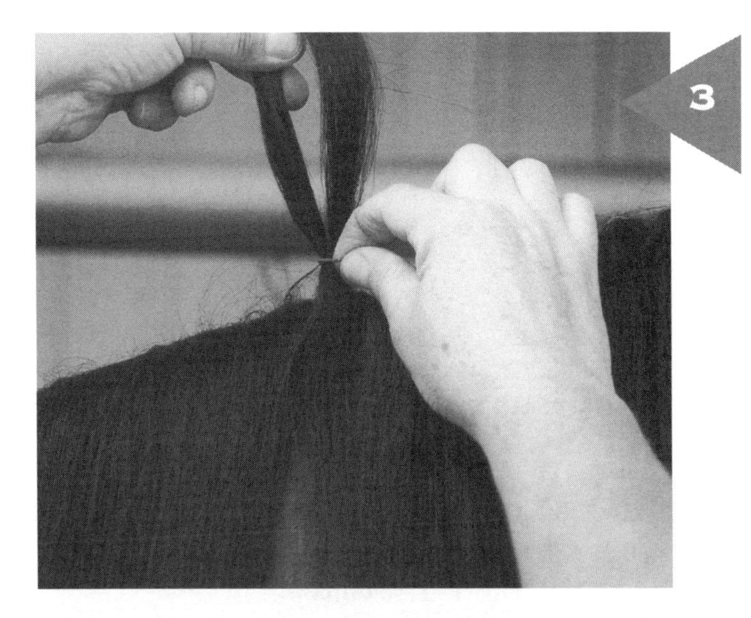

3

Pull the end through. You may have to repeat this several times until you feel that the rubber band is tight enough to hold the hair in place, but not so tight as to cause discomfort to the horse.

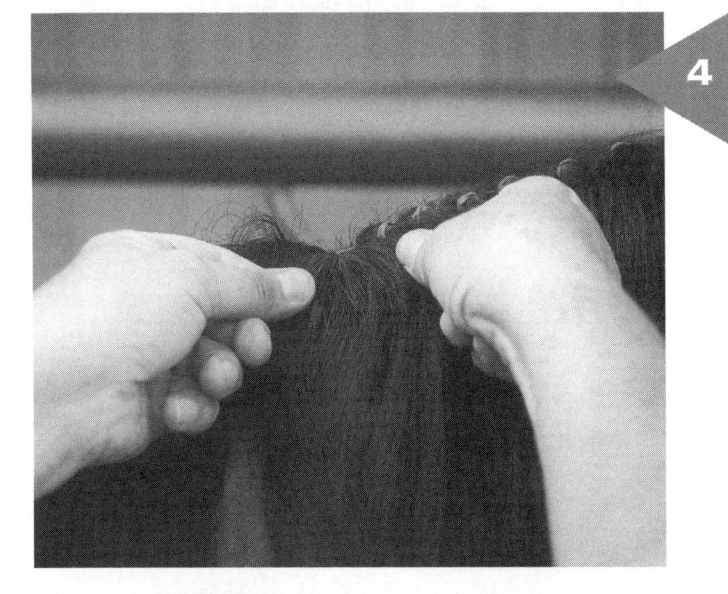

4

Snug it up to the crest, but not too snug. Tight bands will be uncomfortable and make the horse rub his neck in an attempt to get rid of the rubber bands.

A Long Mane Pasture Banded

Here is a 2-year-old with his long mane pasture banded. When he is turned out, it will keep the mane on one side so it will be less likely to become tangled. If you want to band a short mane, the technique is the same. You can even purchase banding tools and rubber bands that will (after you master the technique) make the finished job uniform.

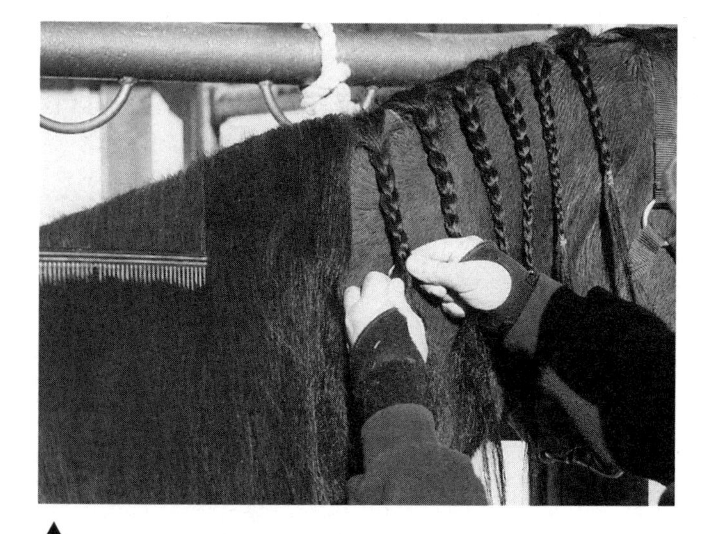

▲
THREE-STRAND BRAID
A more positive and long-lasting means of training a mane and protecting it from tangling is braiding. Section off the mane into a dozen or so little bundles and braid each using the standard three-strand braiding technique. The key is to start the braid loose so it remains comfortable for the horse. If you begin the braid tight right up at the crest, it can drive the horse into an itching frenzy.

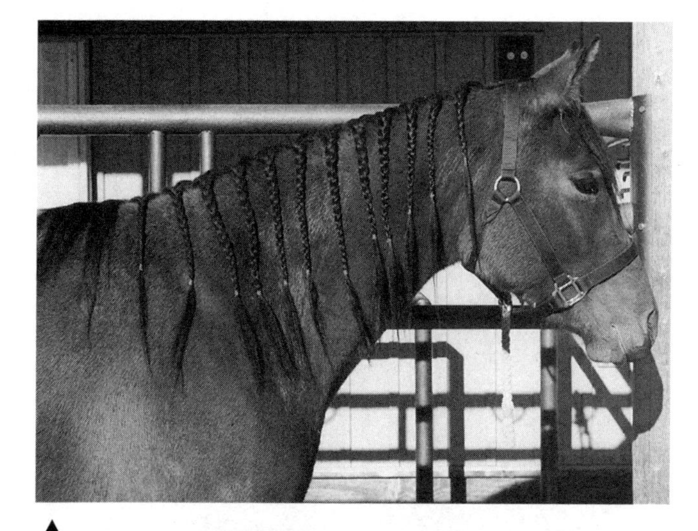

▲
BRAIDS DON'T TANGLE
This 2-year-old is ready to be turned out. The braids can be left in for several months or more. Although the hair will be wavy when the braids are undone, that will be easy to remedy with a quick shampoo. If the horse is left with a loose, long mane for a few months, it will require hours of work to untangle.

MANE STYLES

There are many styles of manes, according to breed and use. Five different ones are shown here.

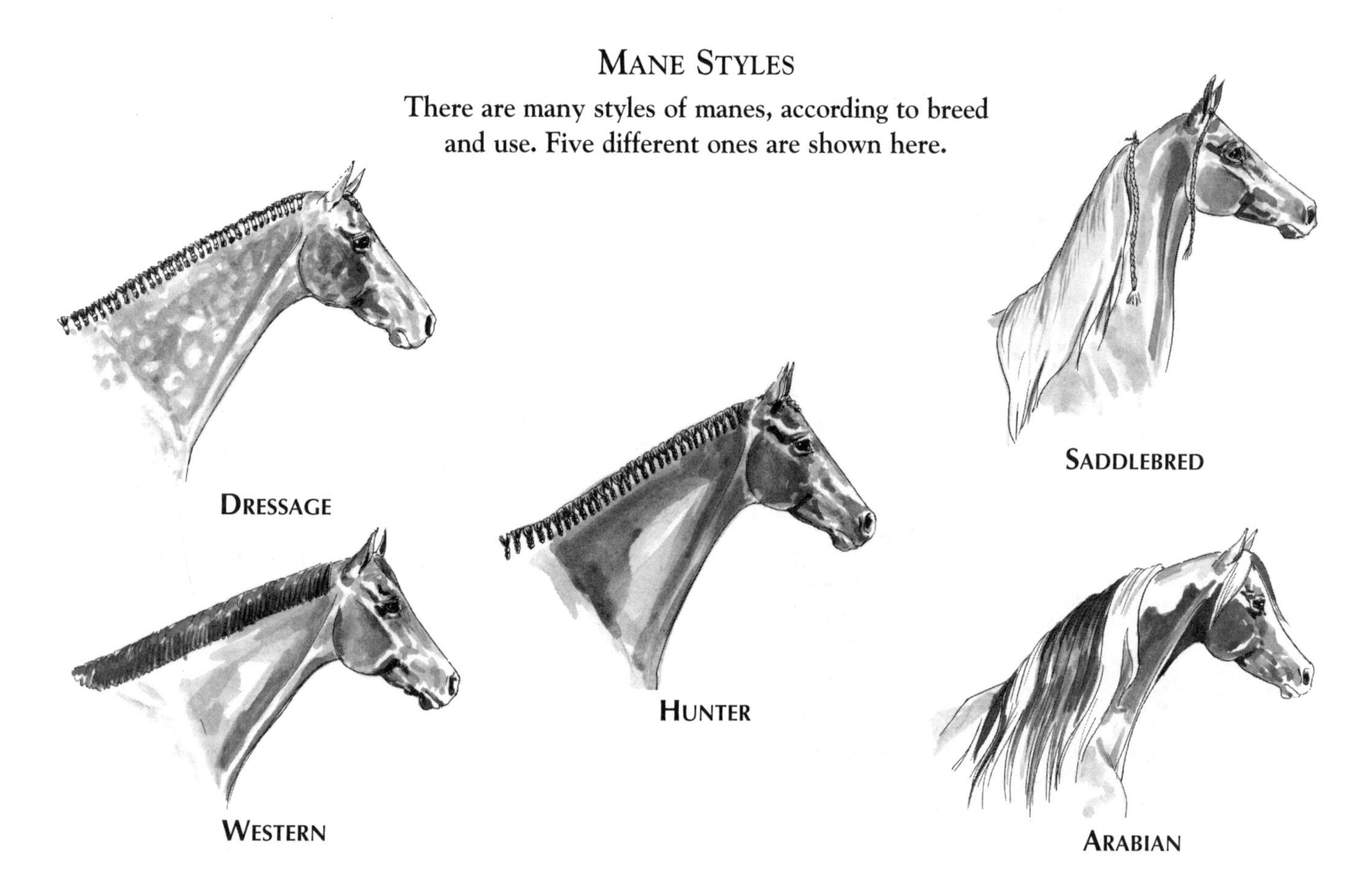

Braiding the Mane for Show

If you want to braid your horse's mane for hunter or dressage classes, follow these steps.

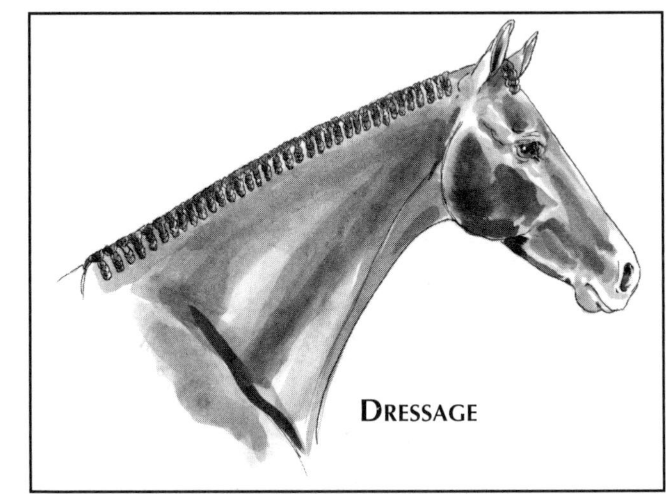

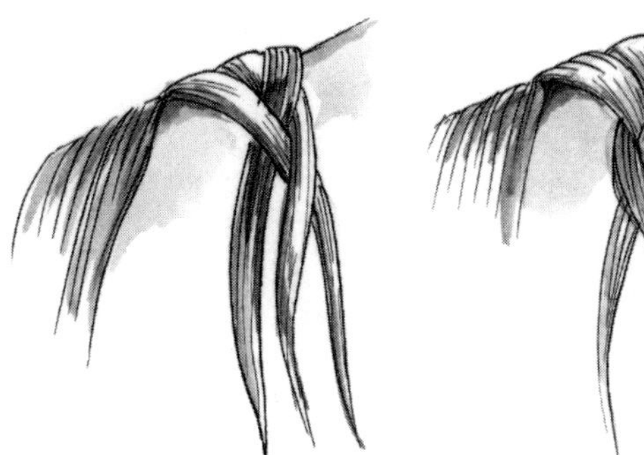

1. Start a three-strand braid, wrapping right over center.

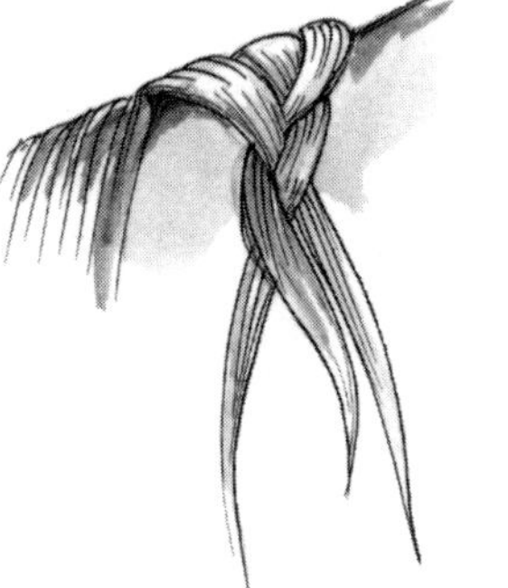

2. Continue by crossing left over center.

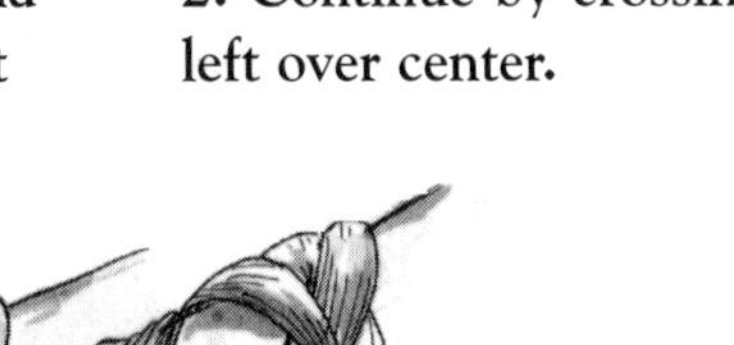

3. Continue to ends of hair.

4. To secure the end of the braid, place a folded string behind the braid.

5. Wrap the string around the braid and through the loop at the fold in the string.

6. Lace the string through the top of the braid and fold the braid in half.

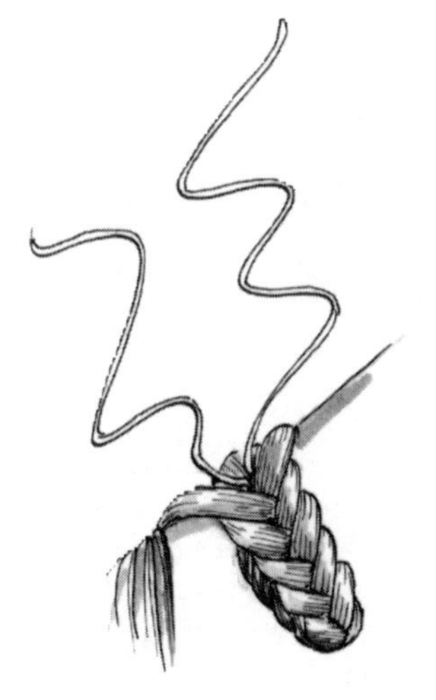

7. Separate the ends of the string that you pulled through the top of the braid.

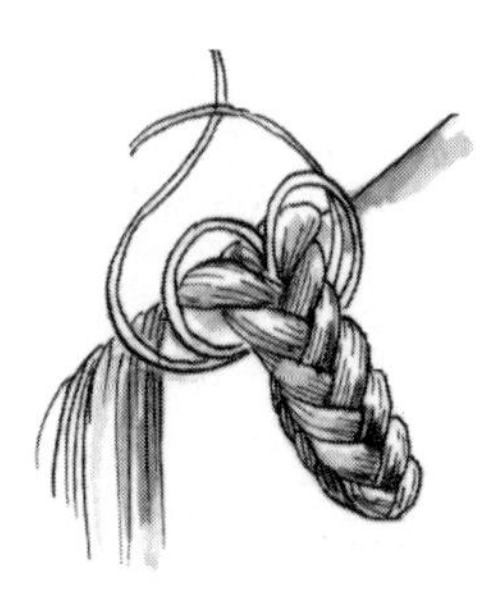

8. Encircle the top of the braid with thread.

9. Secure with a square knot.

Caring for the Tail

Tail Styles

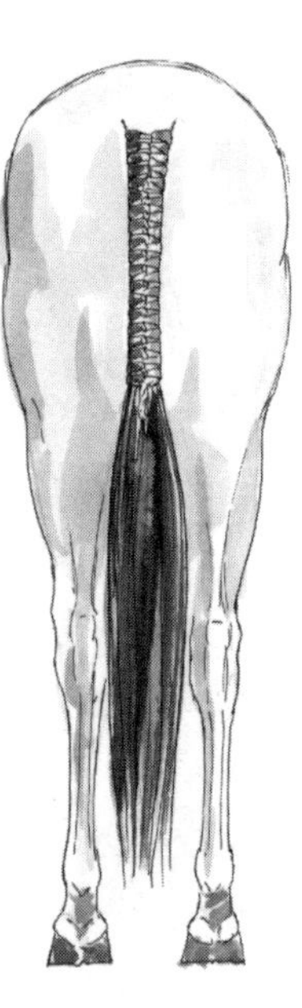

Braided Tail
Hunter

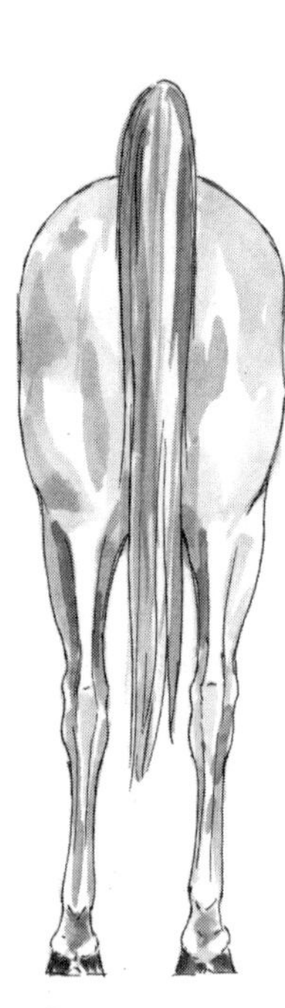

Saddlebred

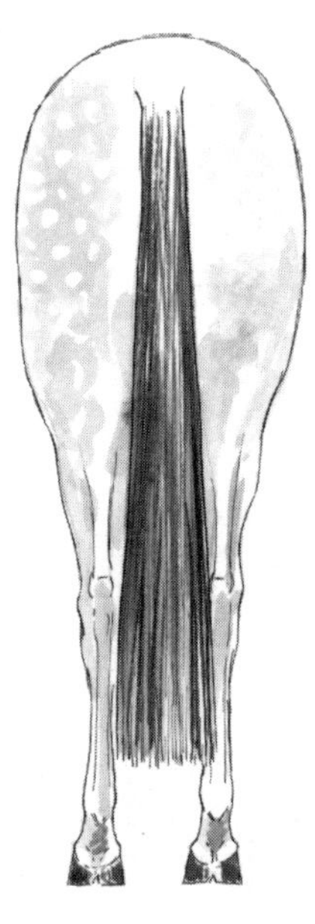

Banged Tail
Dressage

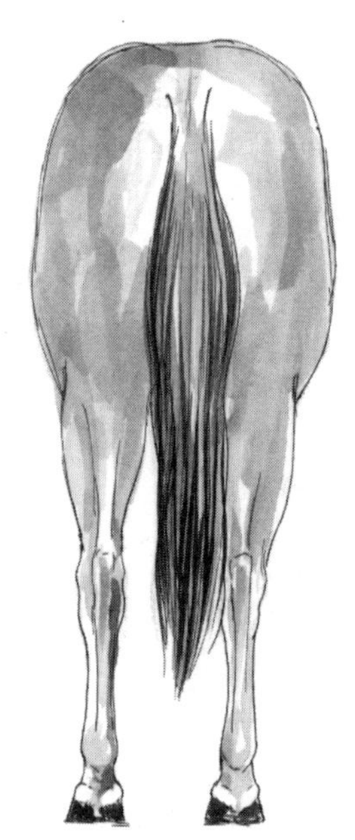

Tapered Tail
Stock Horse

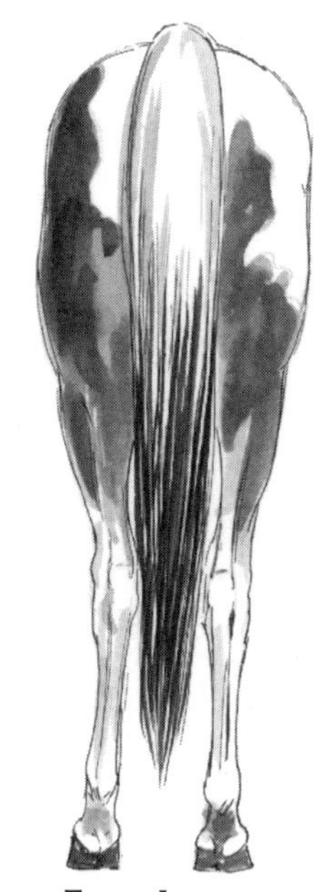

Full-Length
Natural Tail, Arabian

Braiding the Tail for Show

The diagram gives instructions if you want to braid your horse's tail for a hunter show.

1. Dampen around the dock.

2. Separate three small sections of hair on either side.

3. Begin a three-strand braid with those small sections.

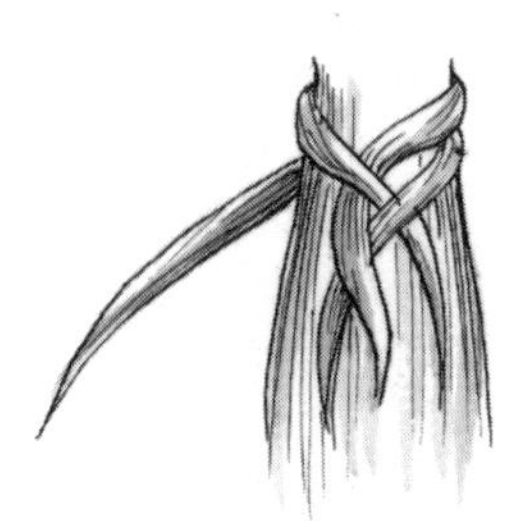

4. Bring each new strand from behind the tail.

5. Add some strands from behind the tail at each step.

6. Keep tension even as you braid.

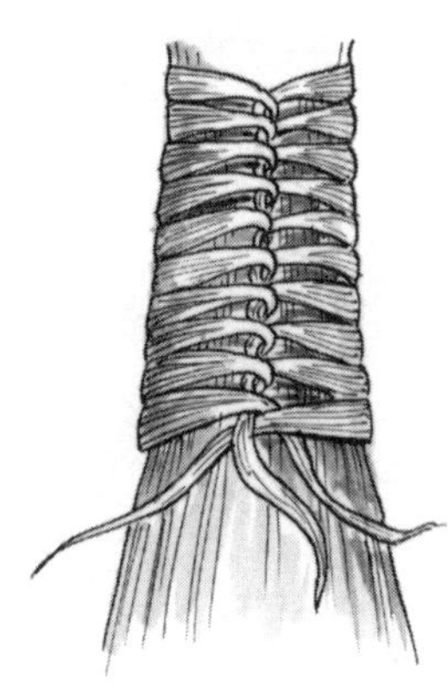

7. Stop adding new strands at the end of the tail bone.

8. Secure the end with a small rubber band.

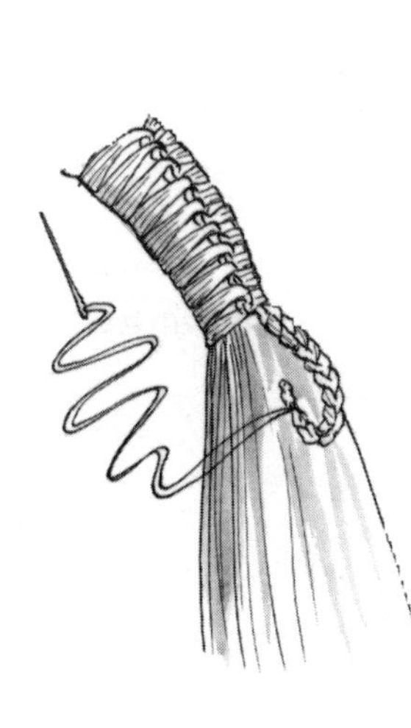

9. Sew through the end of the braid with a needle and yarn.

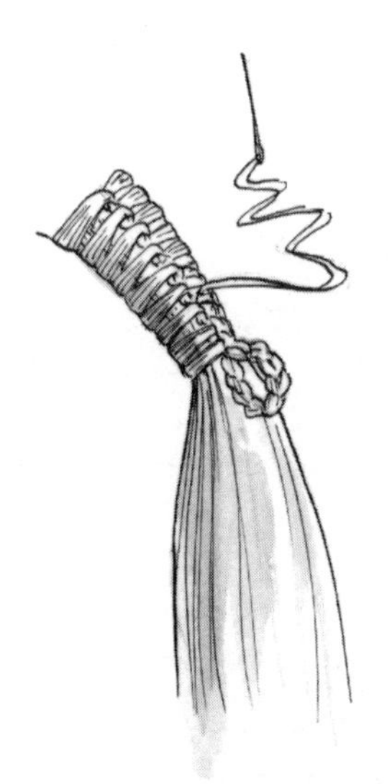

10. Draw the braid into a loop by bringing the needle and yarn out at about the fourth or fifth braid segment. Tie the yarn off in a square knot and snip.

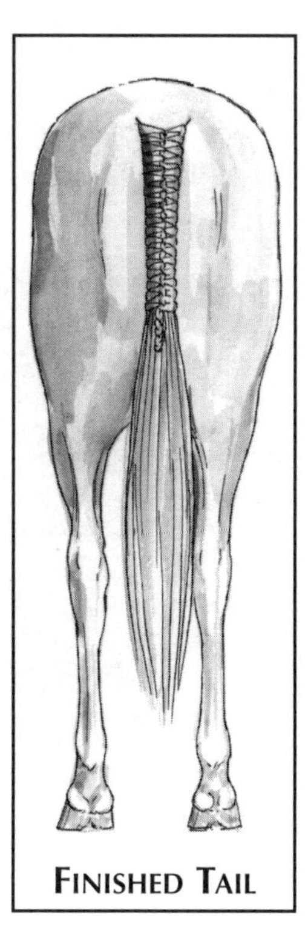

Finished Tail

BANGING A TAIL

One of the simplest and most widely applicable methods for shaping tails is banging the tail, which means cutting the bottom of the tail straight across at fetlock level. The blunt cut looks tidy and prevents the horse from stepping on long hairs and pulling out tail sections.

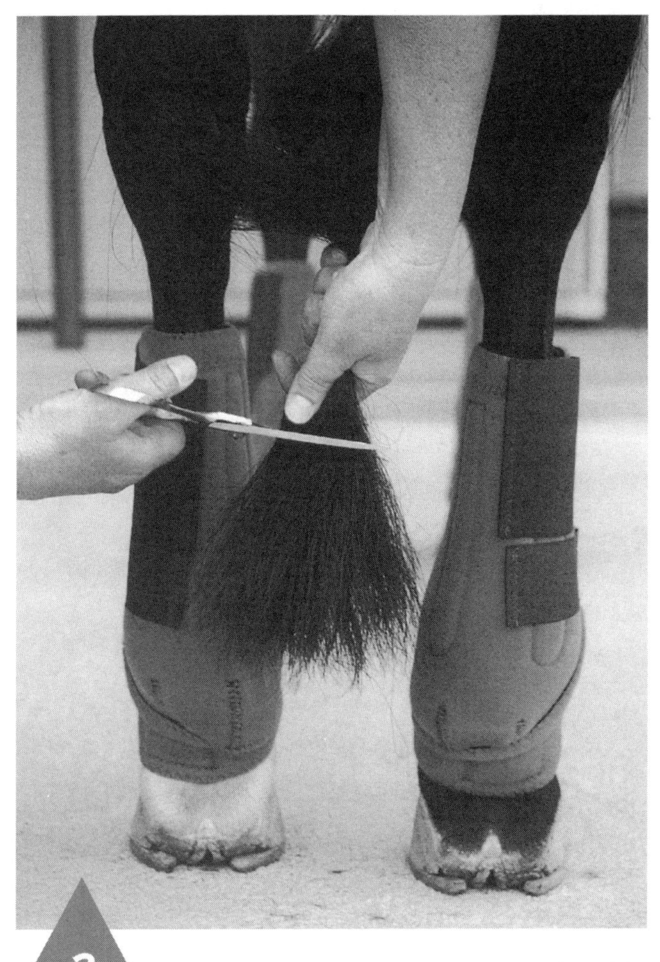

1 Begin with a clean, thoroughly brushed tail. Stand the horse relatively square on a level surface, grab the tail in a bunch at about hock level, and establish how much will have to be removed to bring the tail up to fetlock level.

2 Raise up the tail to a more comfortable cutting level for you. Holding the shears horizontally, cut off the 3 inches that you determined needed to be removed.

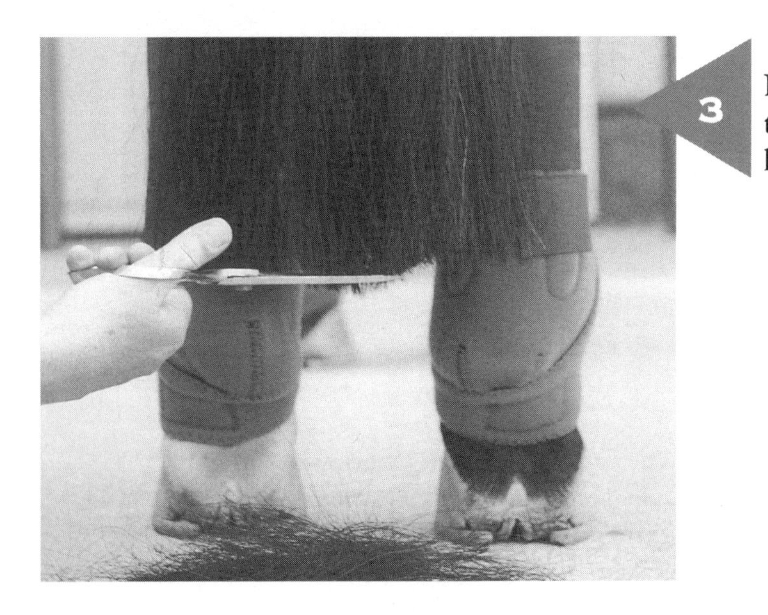

3 Let the tail back down and brush it out again. Check to see that your cut was level and that there are no long stragglers.

BRAIDING A TAIL

Another way to slightly shorten your horse's tail and protect it is to braid it from the end of the tail bone down. This is a good way to protect your horse's tail during non–fly season.

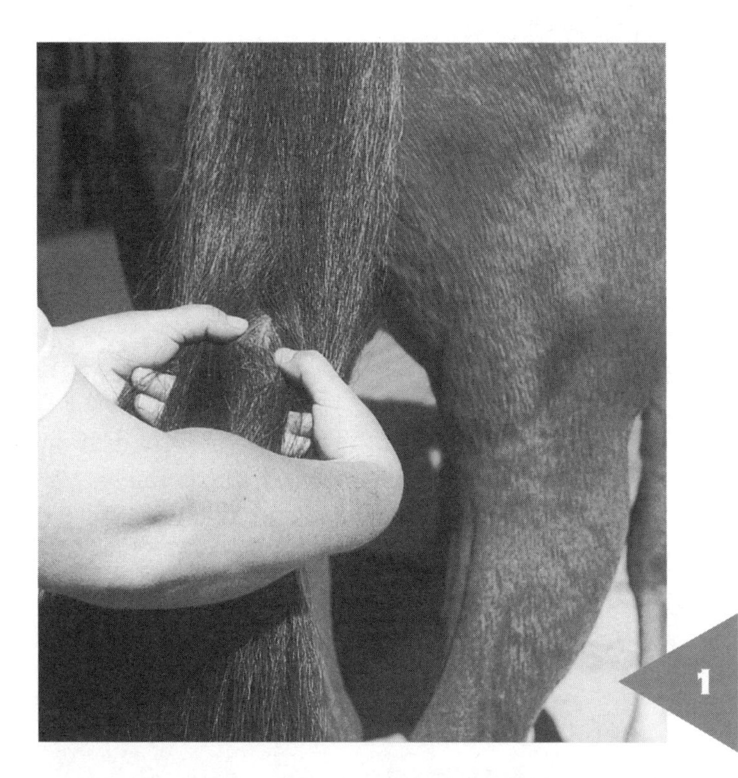

1 First locate the end of the tail bone. On this horse it is quite low. This is the point where you will begin your braid.

2 Separate the tail into three equal sections to begin a three-strand braid.

3 Braid using moderate but even tension.

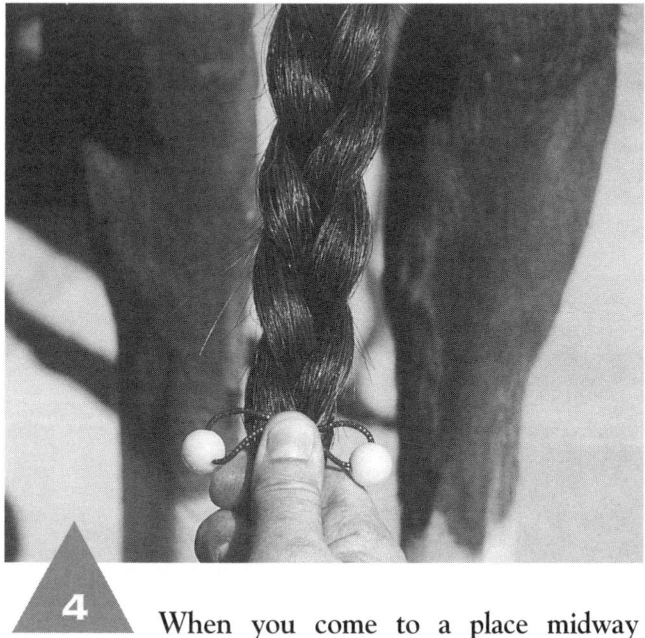

4 When you come to a place midway between the hocks and fetlocks, stop braiding. Take a double-ball ponytail elastic and place the middle of the elastic under the thumb of your right hand and on top of the braid where you stopped braiding.

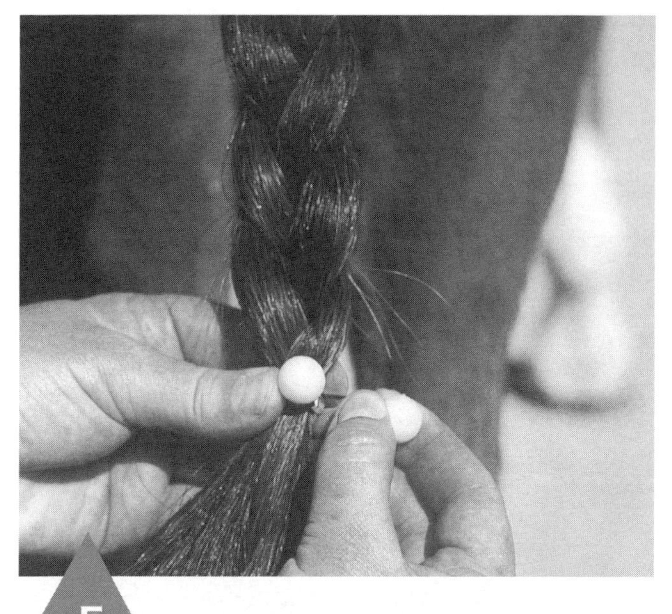

5 With your other hand, wrap the ball on the right around behind the braid and to the spot at the center of the braid on top. Hold it there with your thumb while you take the ball on the left and wrap it around behind the braid, bring it back, and fasten it by looping it over the other ball.

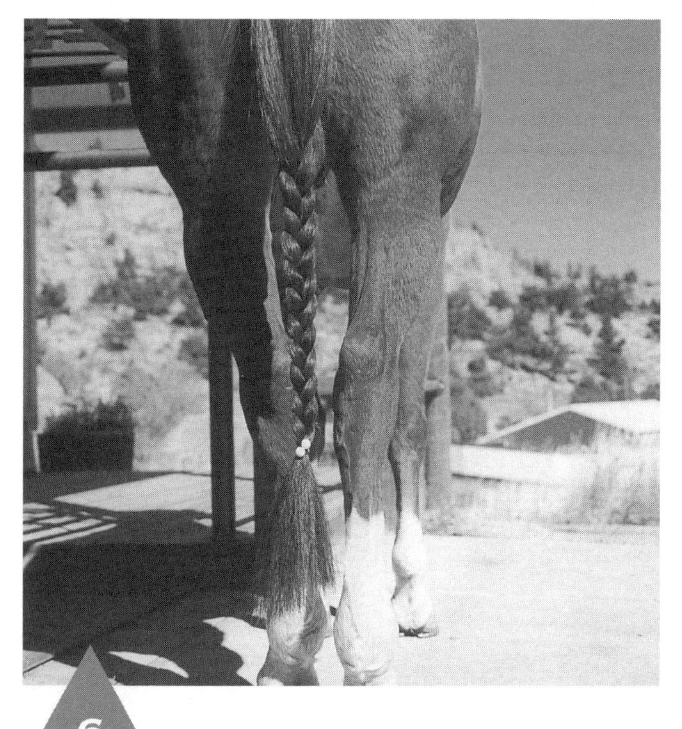

6 Here is a tail ready for non–fly season turnout. The braiding shortens the tail by about 3 inches.

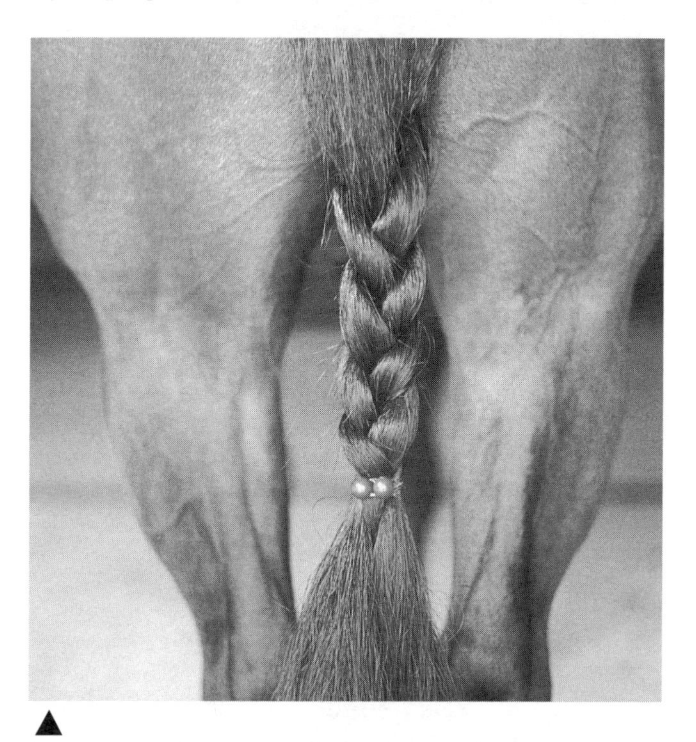

▲

Longer Swatch Can Swat Flies

One way to protect your horse's tail but still give him some swatting power (such as when there are just a few flies around) is to make a much shorter braid and leave a much longer swatch at the end. This way the horse can still reach his sides with his tail.

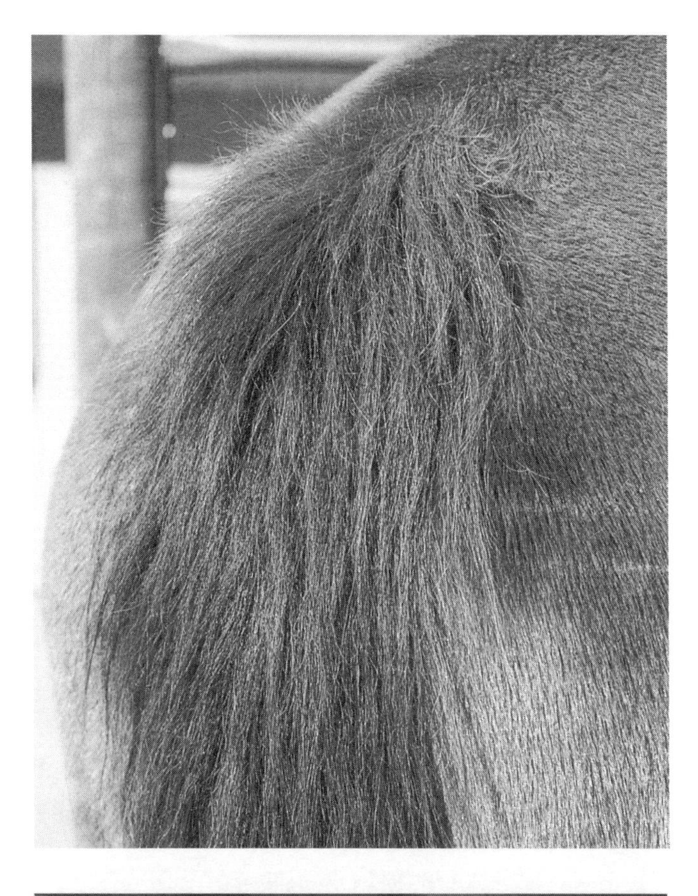

◀ AN UNRULY TAIL HEAD

Often a horse's tail head becomes a bushy mess. If you are going to show in dressage, there are ways you can clip a tail head like this to tidy it. But if you will be showing in Western classes or in hunter classes that require braided tails, you won't want to clip the top of the tail because then you won't be able to braid it. So what you can do is periodically braid the top of the tail to train it to lay more neatly. This is similar to the way in which you would prepare your horse's tail for a hunter show, but remember — this is just for at-home tail training.

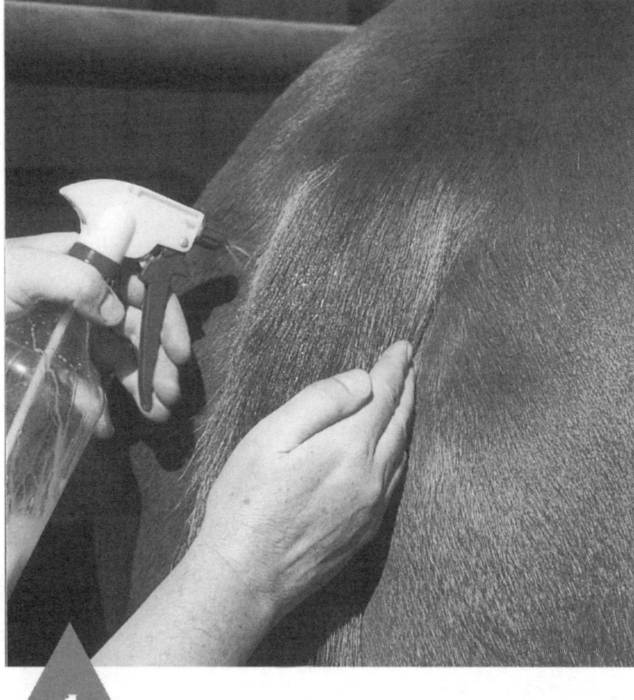

1 Wet the top of the tail with water or a tail conditioner, but don't use anything that will make the hair slippery or you'll have a tough time hanging on to it while you braid. Comb the tail head area and spray some more. Get the wet hairs to lay flat before you begin.

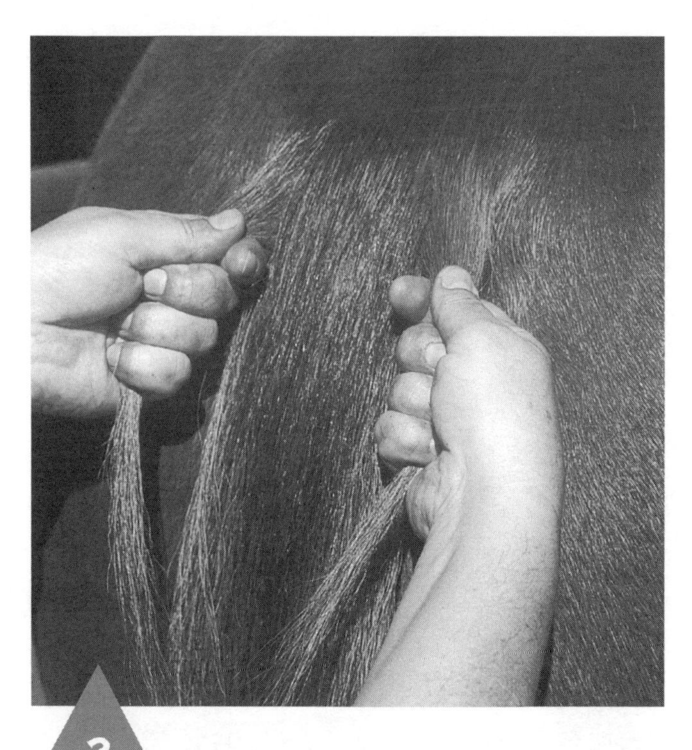

2 Reach around to the underside of the horse's tail and find the border between the tail skin of the dock and the place where the tail hairs grow. Take one strand of hair from the very top of each side of this border.

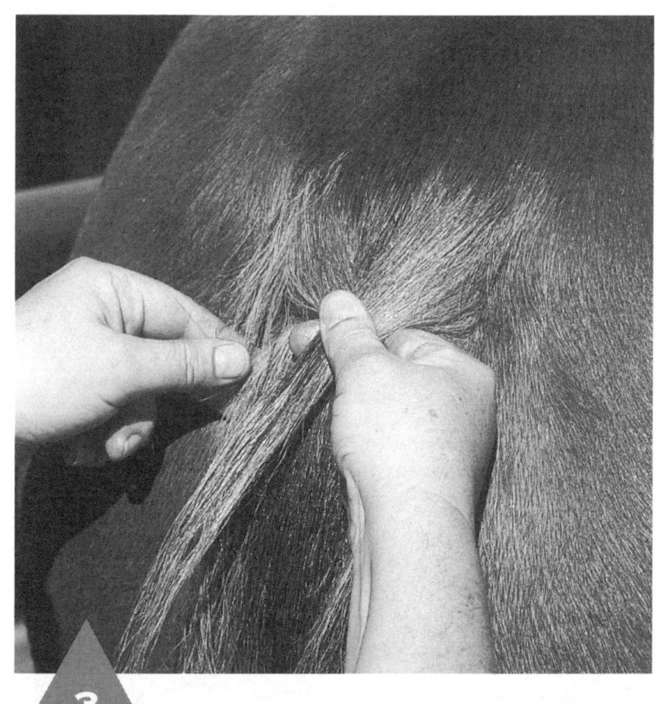

3 Cross right over left and hold the result with your right hand.

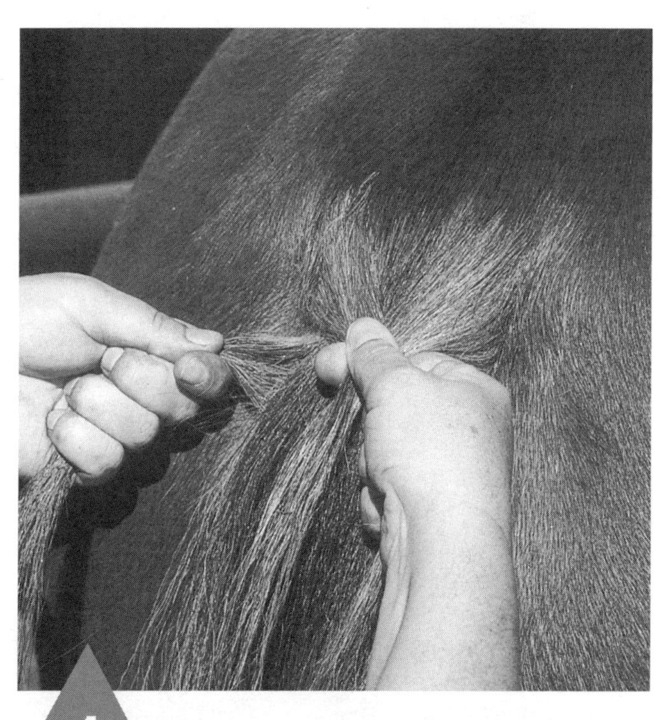

4 Use your left hand to take another strand from the left side. This makes the third strand for your three-strand braid.

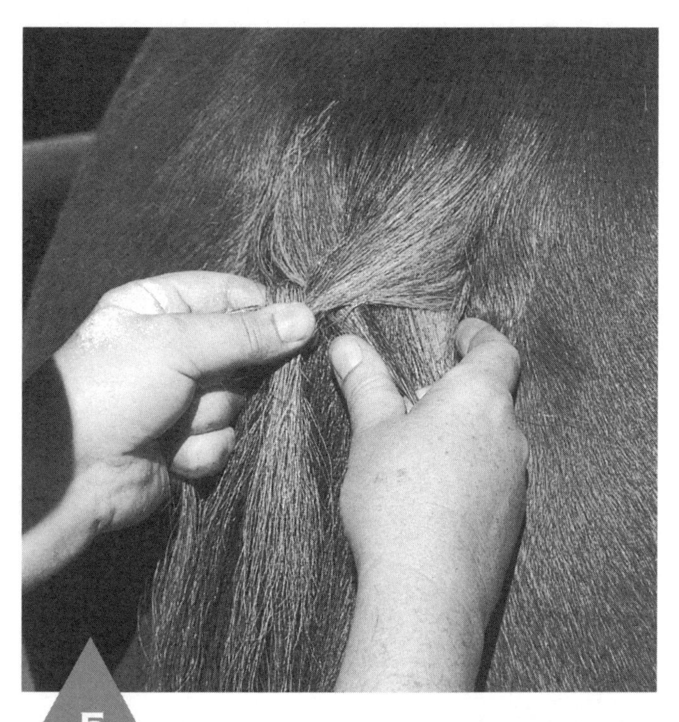

5 Now cross the new strand over the top strand in your right hand and hold those two in your left hand, separated by fingers. Take the bottom strand from the original pair in your right hand.

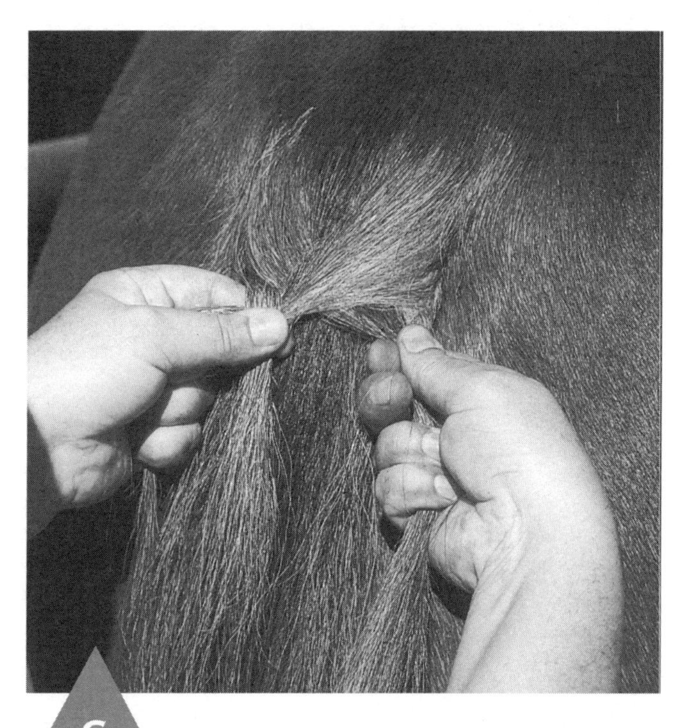

6 Add to it another section from the right side of the horse's tail. Cross what you have in your right hand over the top strand in your left hand and hold those two in your right hand. Take the bottom strand in your left hand and add to it another strand from the left side of the horse's tail.

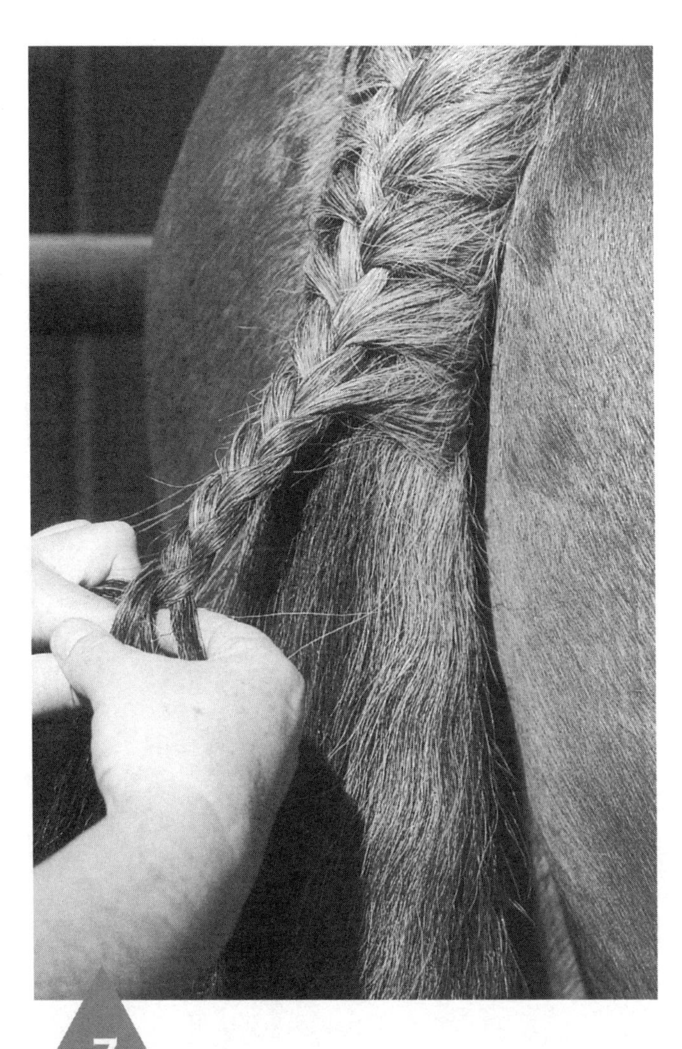

7 Continue until you have been adding hair from the sides for about six or seven sequences or about 8 inches. Stop adding from the sides and just finish off by braiding a regular three-strand braid without adding any more tail hair.

8 This is the finished at-home, tail-training braid. Don't give in to the temptation of snipping the short hairs that try to sneak out of the top of the braid. If you do, you will always have to snip them. If you persist in your training efforts, the hair will grow long enough to blend in nicely with the tail top. You can leave the training braid in for a few days, then remove it, wet the tail top, brush it into place, and wrap it. Remove the wrap after a few hours. Don't comb through the top of the tail. Just let it stay in its "set" position for as long as possible to train the hairs.

Putting Up the Tail for the Winter

Putting up a tail for the winter keeps the long hairs from accumulating frozen mud and ice balls and subsequently breaking. If you braid with even, moderate tension —not tight— the braid will stay in yet be comfortable for the horse. I have left tails up for the winter like this for four to six months, and when I take them down in the spring and shampoo the tails, I find that the tails have grown so much that I have to cut 3–6 inches off to bring the tail to fetlock level.

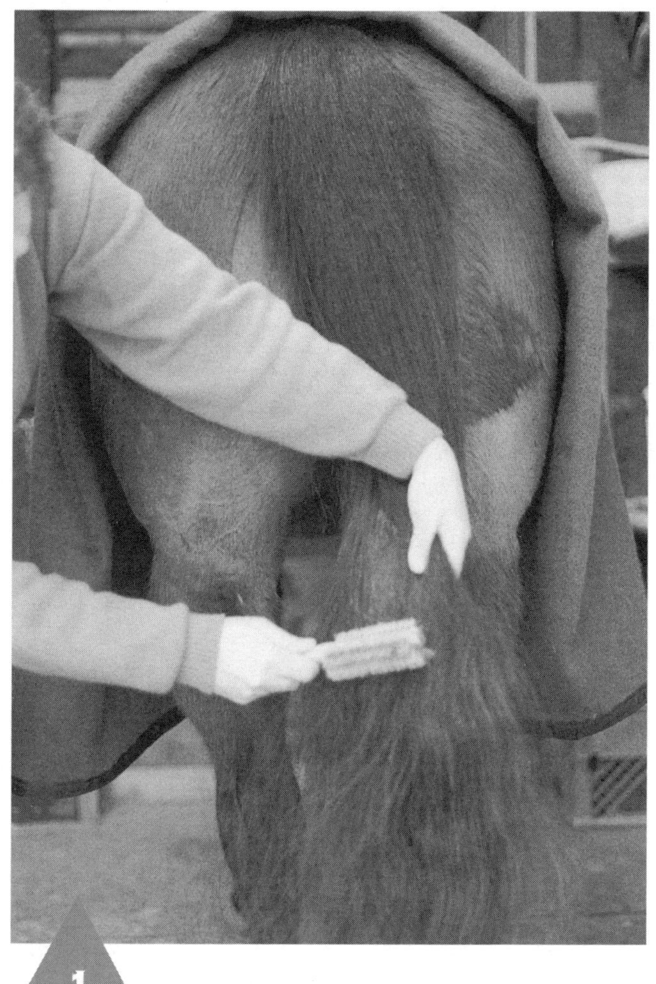

1 Make sure that the tail is thoroughly clean, dry, and brushed before you begin.

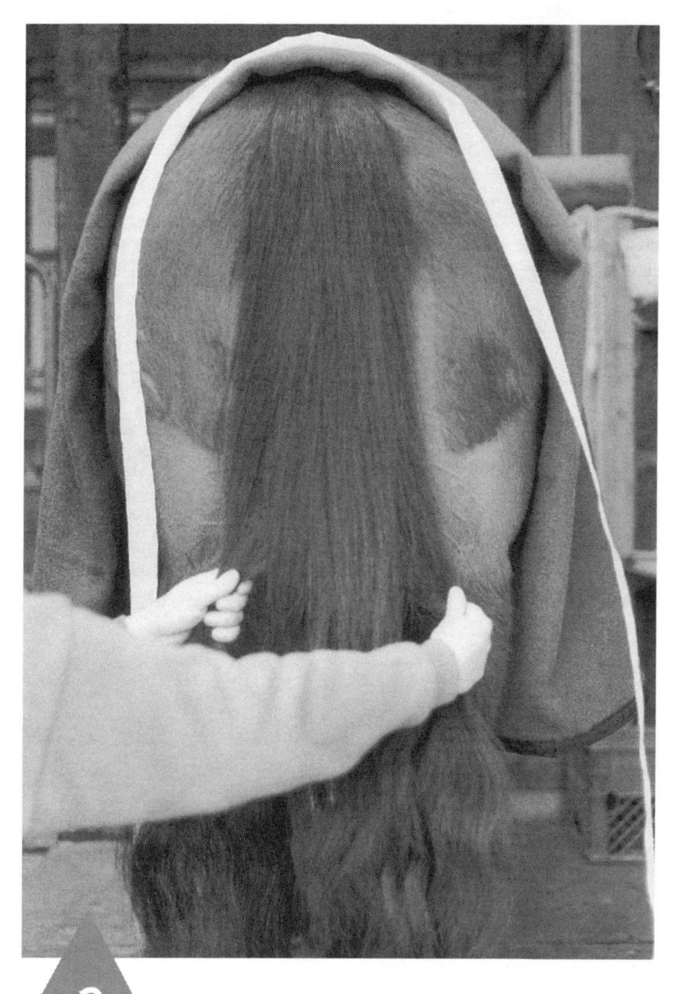

2 Starting at the bottom of the dock, divide the tail into three equal sections.

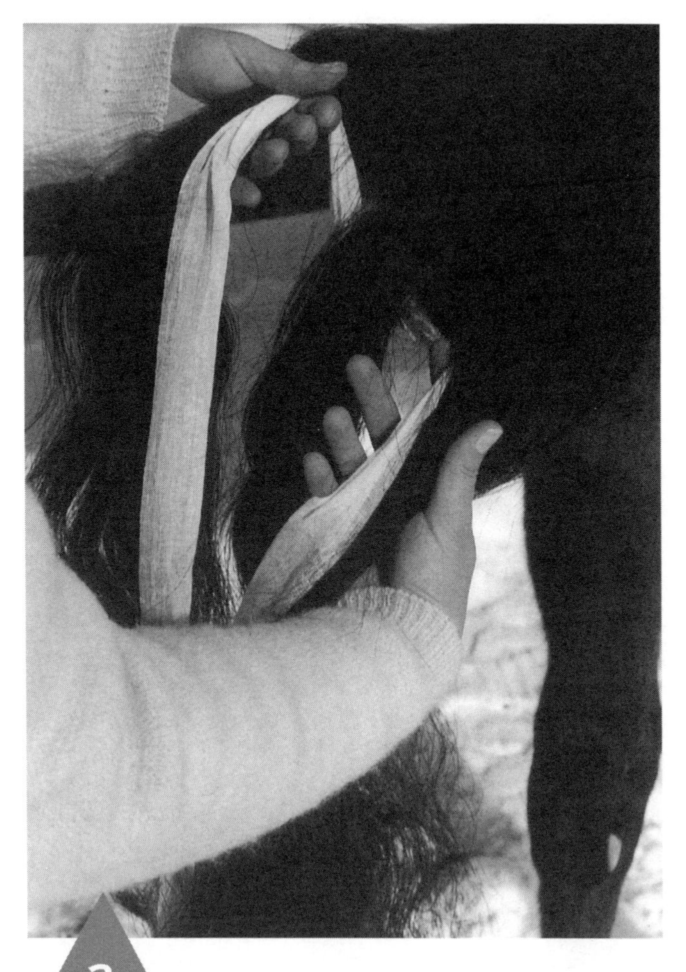

3 Take a 10-foot piece of gauze bandage (¼ inch to 1 inch wide). Place the midpoint of the gauze behind the middle section of tail hair and hold one strand of gauze in each hand along with the other sections.

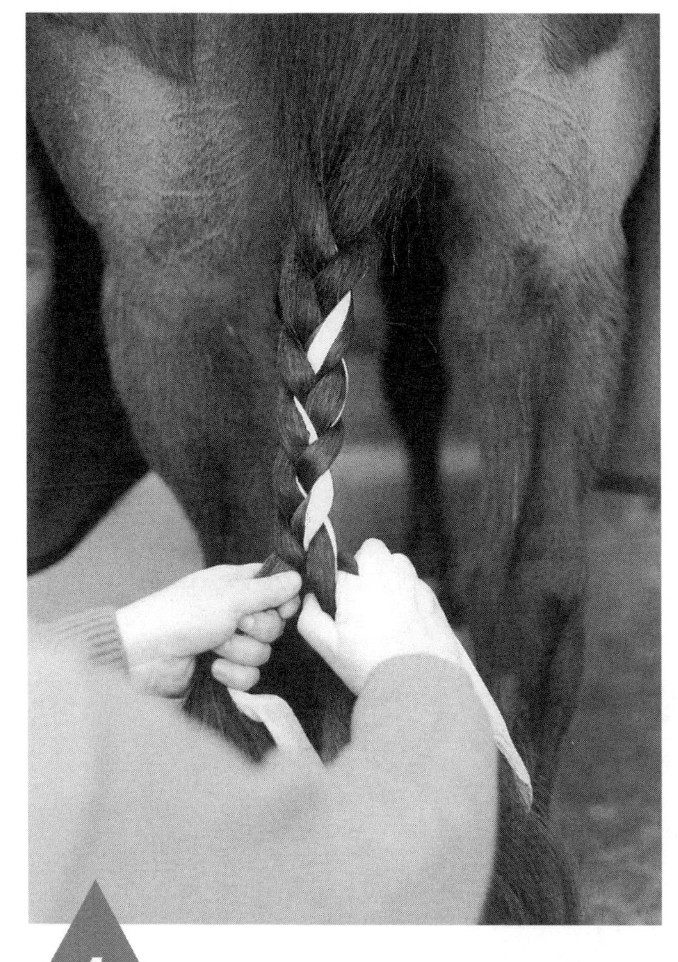

4 Braid the tail as usual, incorporating the gauze into the braid.

5 Stop when you are about 12 inches from the end of the tail. There should be 1 to 2 feet of gauze left on each strand. Using the gauze, tie two overhand knots at the end of the braid.

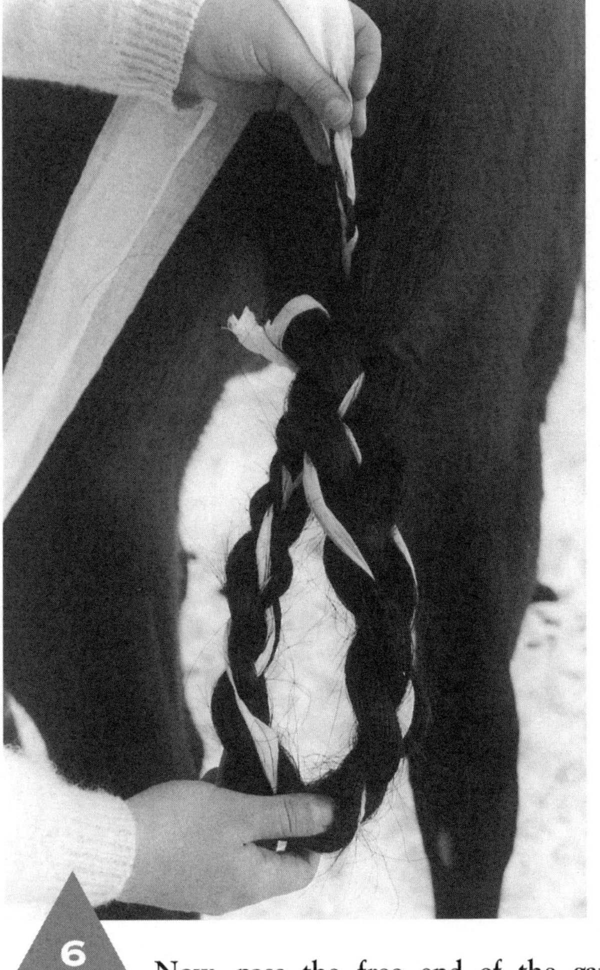

6 Now, pass the free end of the gauze through the tail where the braiding began.

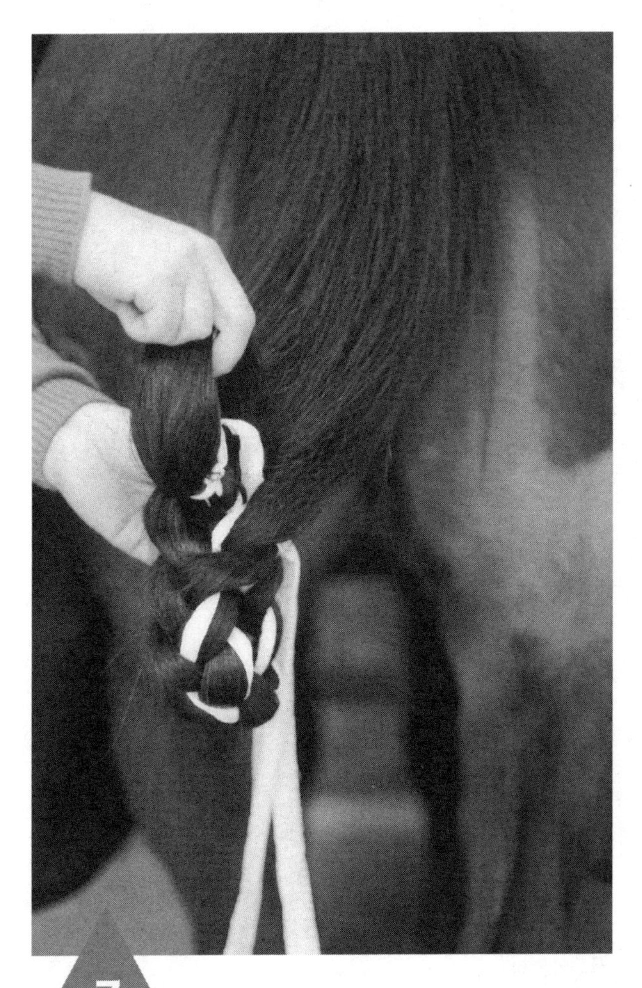

7 It is easiest to do this by poking part of the lower braid through a hole in the top of the braid and then pulling the unbraided tuft and gauze strips through.

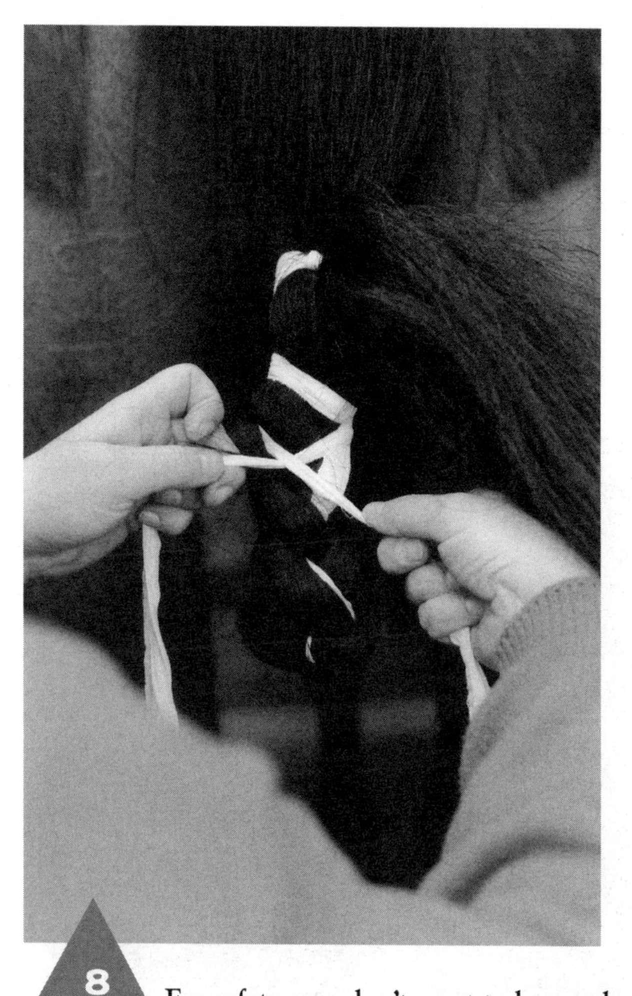

8 For safety, you don't want to leave a loop in the braided tail. Use the gauze strips to secure the two pieces of braided tail together by crisscrossing the gauze before tying it off with two overhand knots.

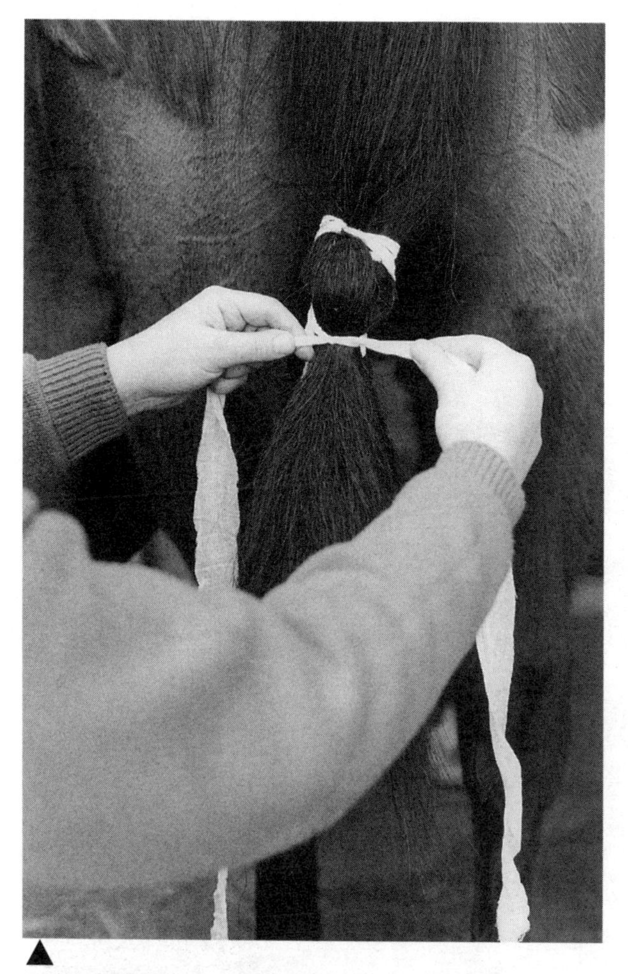

▲

LAY UNBRAIDED SECTION OVER BRAIDED TAIL

Another way of finishing is to lay the unbraided tuft over the top of the crisscrossed braided tail and secure it in two places as shown. This ends up looking like a shorter tail, because you have left a long tuft hanging out of the bottom.

WRAPPING A TAIL

Wrapping a horse's tail is commonly used in the following situations: whenever the horse will be traveling in a trailer, to prevent him from rubbing his tail out on the butt bar or butt chain; whenever a mare will be palpated, washed, or bred; to train any horse's tail top to lie flat for show.

A wrap should be left on only for an hour or so at a time and the wrap should be put on with moderate tension. If tail wrap is put on too tight or is left on too long, it can cut off the circulation to the tissues of the dock and cause the skin and/or tail hairs to slough.

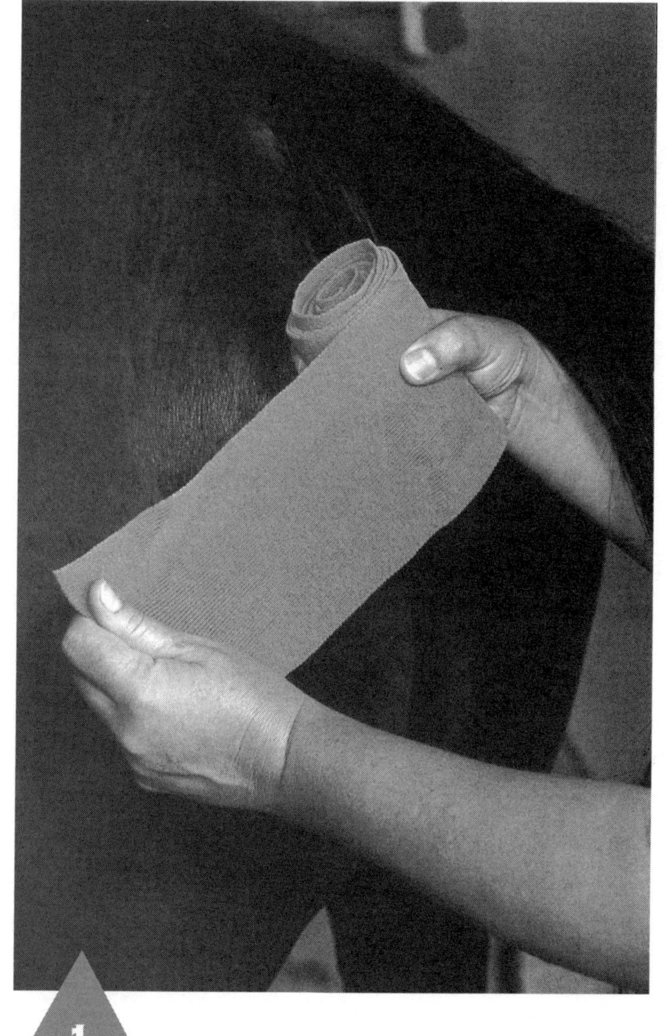

1 Lift the horse's tail and let it rest on your right arm. Hold the track wrap as shown.

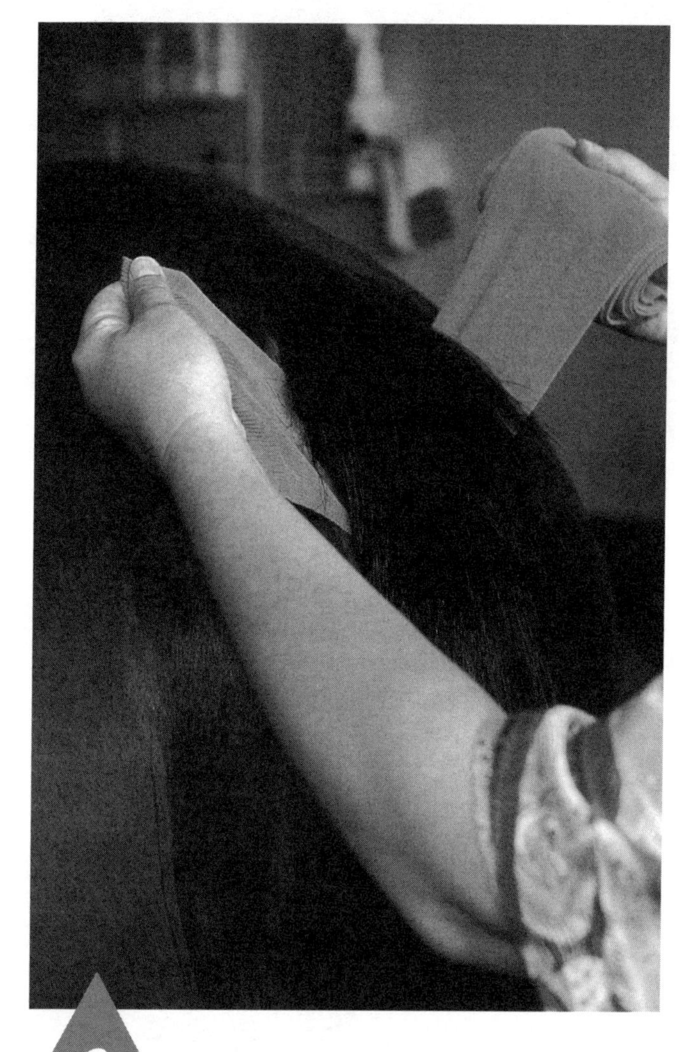

2 Place the wrap under the horse's tail and slide it all the way to the top of the tail. Be careful! A horse who is not used to this procedure might clamp his tail, move away, squat, or even kick when he feels the wrap on the underside of his tail or against his anus.

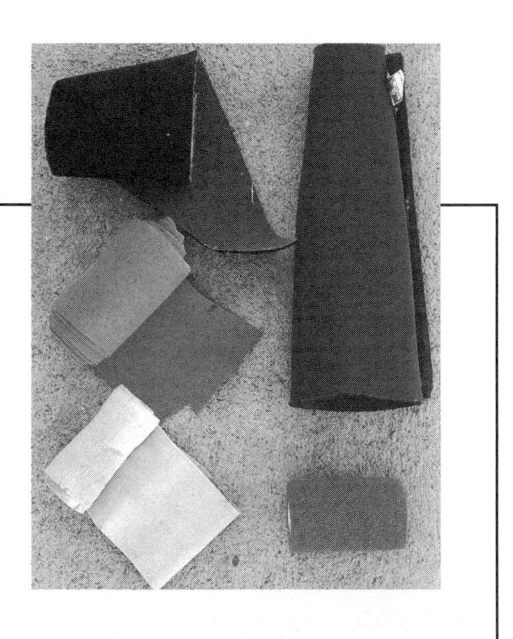

All of these tail wraps are washable and reusable except the last one. Left row from the top: **neoprene bandage** (stretchy, nonslip, nonbreathable, and is often put on too tight; fastens with Velcro), **cotton track bandage** (breathable, tends to slip, is somewhat stretchy, and fastens with tape ties or Velcro), **an Ace bandage** (very elastic and stretchy, so it is often applied too tight; must be fastened with pins or tape). Right row from the top: **neoprene tube-type tail wrap** (this may or may not fit your particular horse's tail; fastens with Velcro), **crepe bandaging material** (stretchy, nonbreathable, disposable, ideal for short-term uses, such as during breeding).

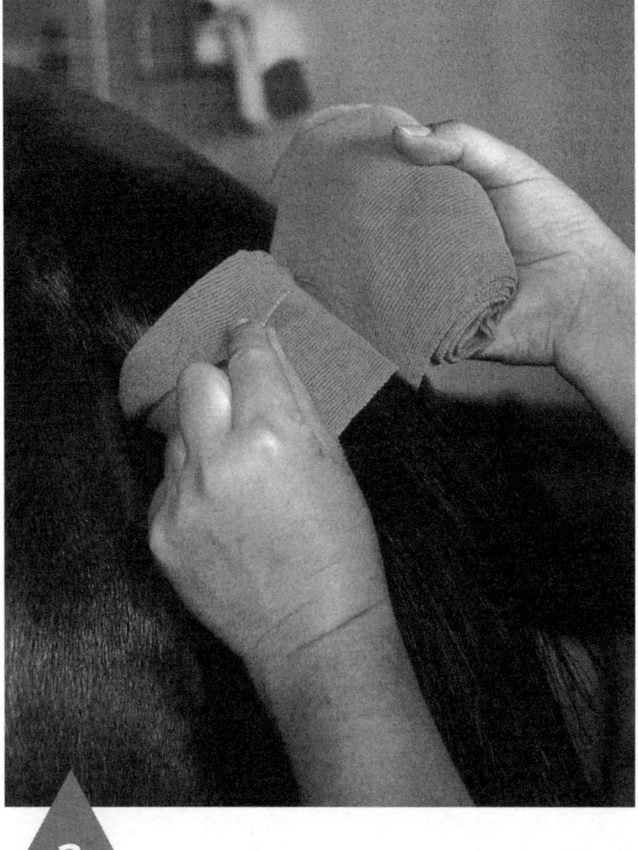

3 Begin wrapping. You can fold the end of the bandage down and cover it with your next wrap.

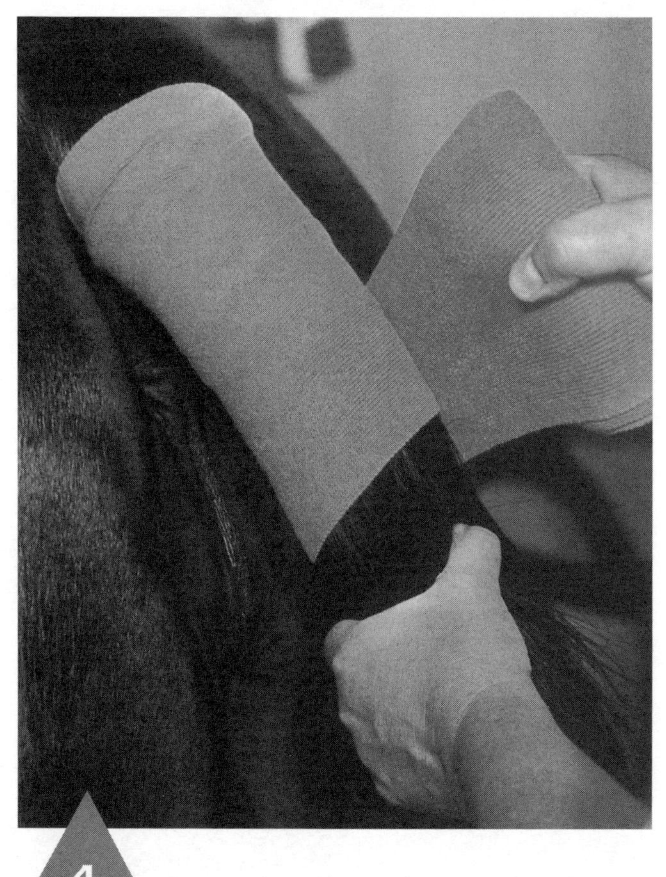

4 Continue wrapping downward with consistent pressure.

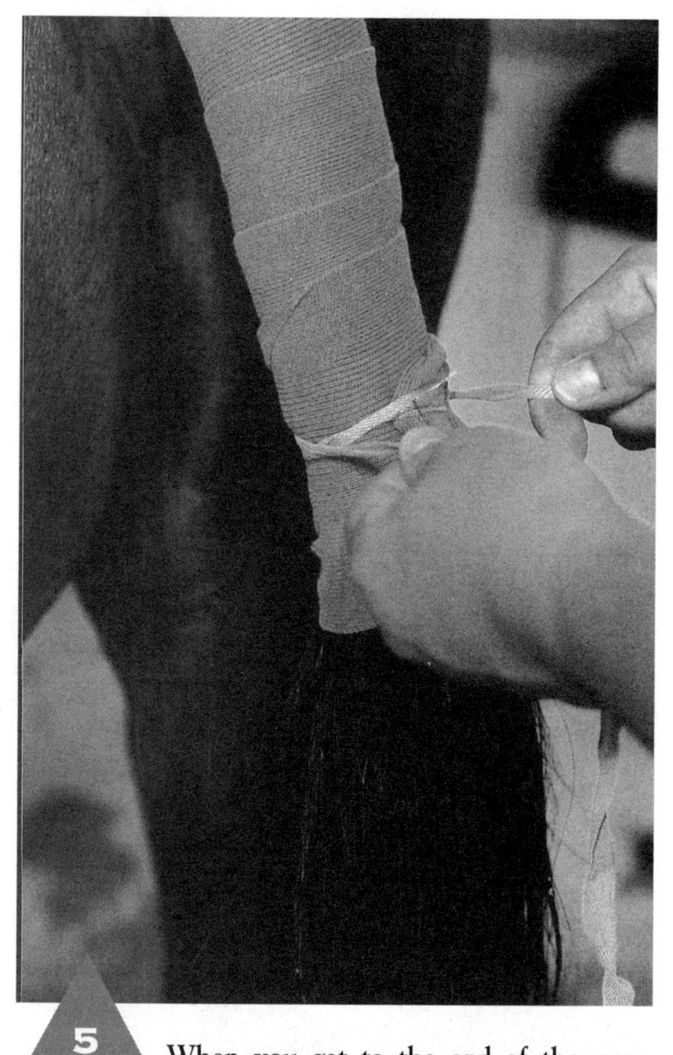

5 When you get to the end of the wrap (which should end just below the end of the dock), take the tape ties to the back, exchange them in your hands, and bring them around to the top of the wrap.

6 Tie an overhand knot and a bow. If you are going to work on the horse, such as for breeding, you don't necessarily need to tuck in the bow and tails, but if you plan to leave the horse unattended, you will want to tuck all the loose ends under the last wrap of the bandage.

Removing the Tail Wrap

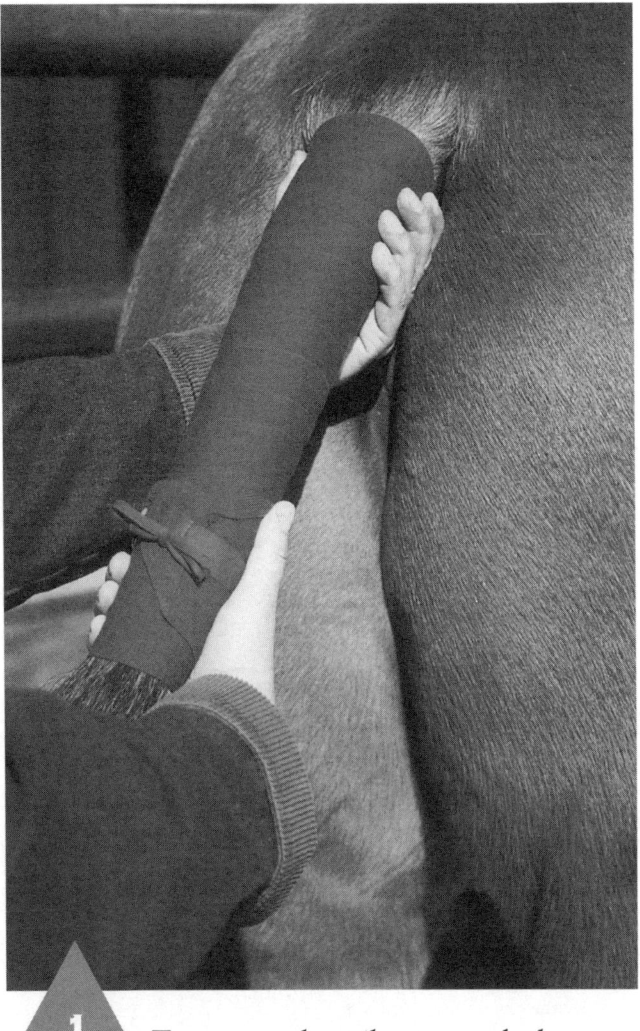

1 To remove the tail wrap, grab the wrap with both hands in this fashion.

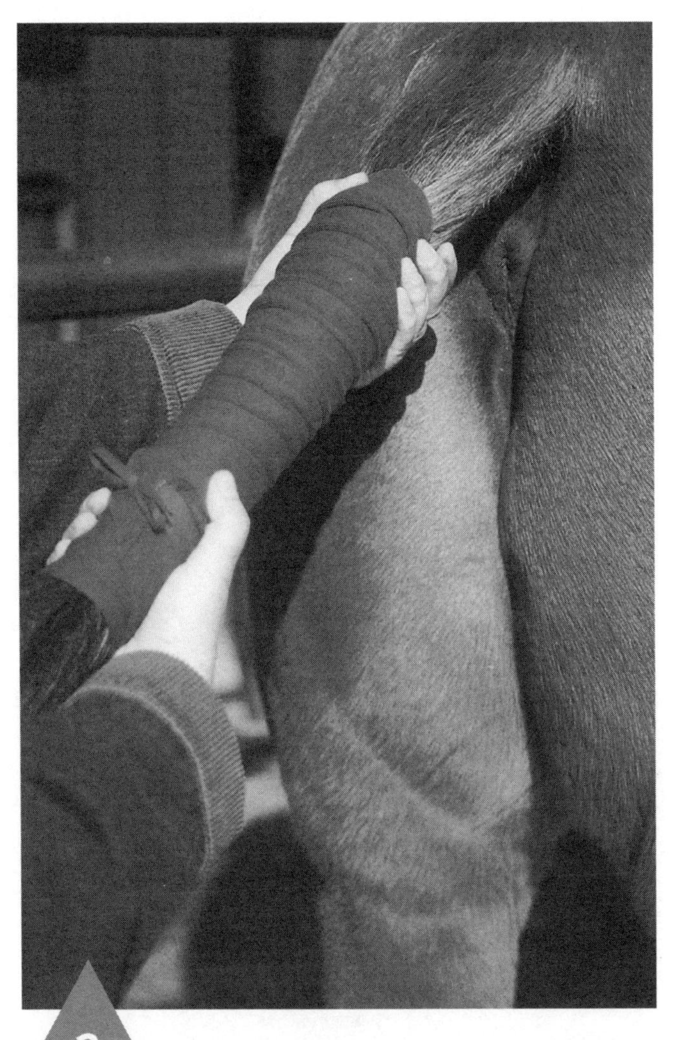

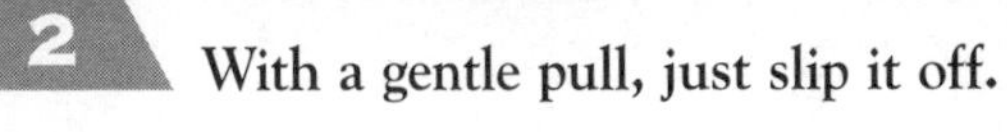

2 With a gentle pull, just slip it off.

Appendix

PARTS OF THE HORSE

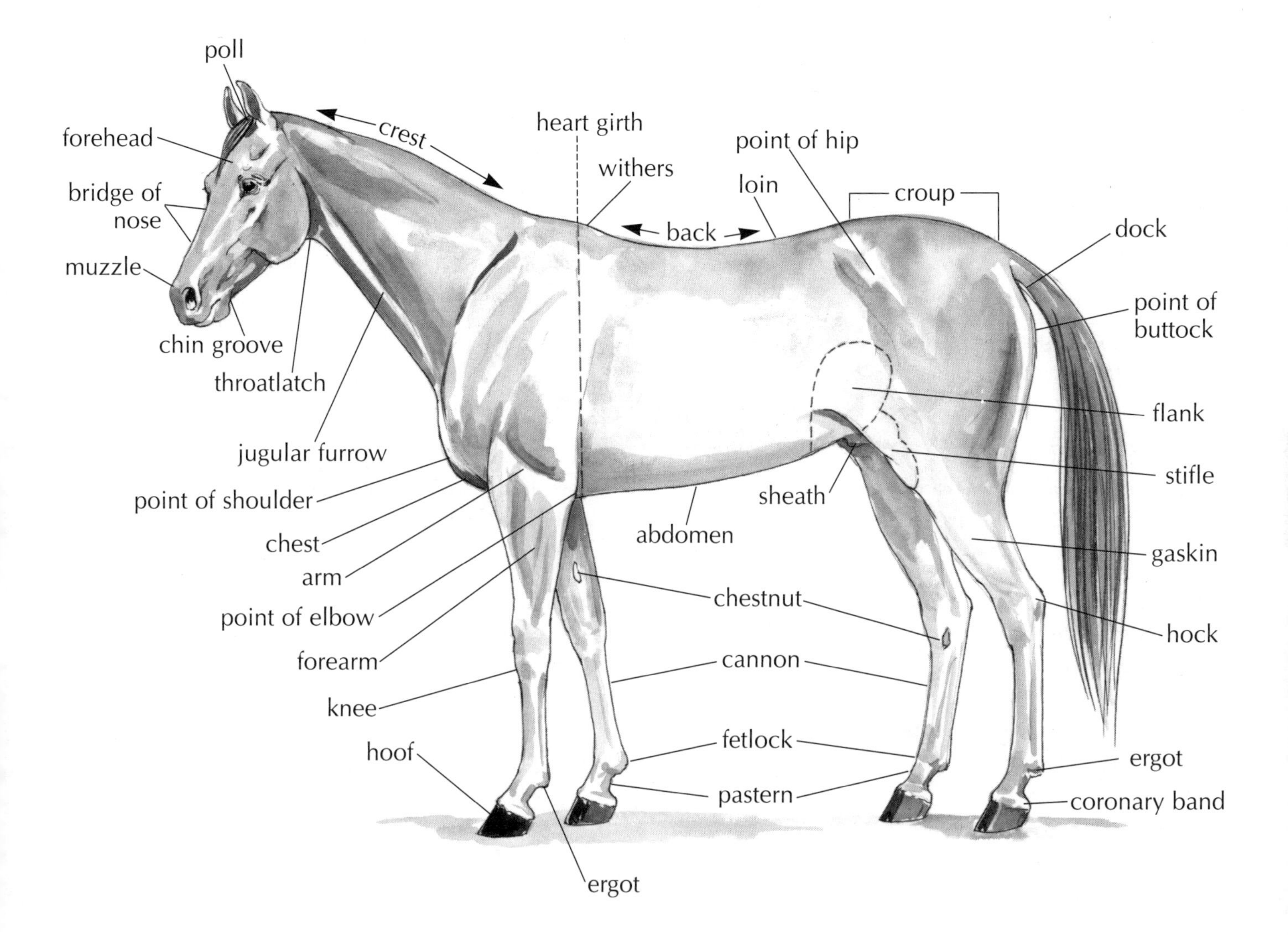

Recommended Reading

Haas, Jessie. *Safe Horse, Safe Rider: A Young Rider's Guide to Responsible Horsekeeping*. North Adams, MA: Storey Publishing, 1994.

Hill, Cherry. *101 Arena Exercises: A Ringside Guide for Horse and Rider*. North Adams, MA: Storey Publishing, 1995.

———. *Becoming an Effective Rider: Developing Your Mind and Body for Balance and Unity*. North Adams, MA: Storey Publishing, 1991.

———. *The Formative Years: Raising and Training the Horse from Birth to Two Years*. Ossining, NY: Breakthrough, 1988.

———. *From the Center of the Ring: An Inside View of Horse Competitions*. Pownal, VT: Storey Books, 1988.

———. *Horse Health Care: A Step-by-Step Photographic Guide to Mastering over 100 Horsekeeping Skills*. North Adams, MA: Storey Publishing, 1997.

———. *Horse for Sale: How to Buy a Horse or Sell the One You Have*. New York: Howell Book House, 1995.

———. *Horsekeeping on a Small Acreage: Facilities Design and Management*. North Adams, MA: Storey Publishing, 2005.

———. *Making Not Breaking: The First Year Under Saddle*. Ossining, NY: Breakthrough, 1992.

———. *Your Pony, Your Horse: A Kid's Guide to Care and Enjoyment*. North Adams, MA: Storey Publishing, 1995.

Hill, Cherry, and Richard Klimesh, CJF. *Maximum Hoof Power: How to Improve Your Horse's Performance Through Proper Hoof Management*. New York: Howell Book House, 1994.

Kellon, Eleanor, VMD. *Dr. Kellon's Guide to First Aid for Horses*. Ossining, NY: Breakthrough, 1990.

Lewis, Lon. *Feeding and Care of the Horse*, second edition. Baltimore, MD: Williams & Wilkins, 1995.

Stashak, Ted, DVM, and Cherry Hill. *Horseowner's Guide to Lameness*. Baltimore, MD: Williams & Wilkins, 1995.

Page references in *italics* indicate illustrations

Other Storey Titles You Will Enjoy

Cherry Hill's Horse Care for Kids, by Cherry Hill.
The essentials of equine care, from matching the right horse to the rider to handling, grooming, feeding, stabling, and much more.
128 pages. Paper. ISBN-13: 978-1-58017-407-7.
Hardcover with jacket. ISBN-13: 978-1-58017-476-3.

The Foaling Primer, by Cynthia McFarland.
A chronicle of the first year of a horse's life in amazing, up-close photographs and detailed descriptions.
160 pages. Paper. ISBN-13: 978-1-58017-608-8.
Hardcover with jacket. ISBN-13: 978-1-58017-609-5.

The Horse Behavior Problem Solver,
by Jessica Jahiel.
A friendly, question-and-answer sourcebook to teach readers how to interpret problems and develop workable solutions.
352 pages. Paper. ISBN-13: 978-1-58017-524-1.

Horse Health Care, by Cherry Hill.
A complete health care reference of correct techniques for everything from dental exams to leg wrapping and hoof care.
160 pages. Paper. ISBN-13: 978-0-88266-955-7.

Horsekeeping on a Small Acreage, by Cherry Hill.
A thoroughly updated, full-color edition of the author's best-selling classic about how to have efficient operations and happy horses.
320 pages. Paper. ISBN-13: 978-1-58017-535-7.
Hardcover. ISBN-13: 978-1-58017-603-3.

How to Think Like a Horse, by Cherry Hill.
Detailed discussions of how horses think, learn, respond to stimuli, and interpret human behavior — in short, a light on the equine mind.
192 pages. Paper. ISBN-13: 978-1-58017-835-8.
Hardcover. ISBN-13: 978-1-58017-836-5.

Storey's Barn Guide to Horse Health Care + First Aid.
Essential techniques for every horseowner, including dental and hoof care, nutrition, and treating wounds and lameness.
128 pages. Paper with concealed wire-o binding.
ISBN-13: 978-1-58017-639-2.

Storey's Guide to Feeding Horses,
by Melyni Worth.
A complete guide to designing a balanced feeding program according to the individual needs of every horse.
256 pages. Paper. ISBN-13: 978-1-58017-492-3.
Hardcover. ISBN-13: 978-1-58017-496-1.

Storey's Guide to Raising Horses,
by Heather Smith Thomas.
The complete guide to intelligent horsekeeping: how to keep a horse healthy in body and spirit.
512 pages. Paper. ISBN-13: 978-1-58017-127-4.

Storey's Guide to Training Horses,
by Heather Smith Thomas.
Vital information about the training process, written from the standpoint that each horse is unique and needs to learn at its own pace.
512 pages. Paper. ISBN-13: 978-1-58017-467-1.
Hardcover. ISBN-13: 978-1-58017-468-8.

Storey's Illustrated Guide to 96 Horse Breeds of North America, by Judith Dutson.
A comprehensive encyclopedia filled with full-color photography and in-depth profiles on the 96 horse breeds that call North America home.
416 pages. Paper. ISBN-13: 978-1-58017-612-5.
Hardcover with jacket. ISBN-13: 978-1-58017-613-2.